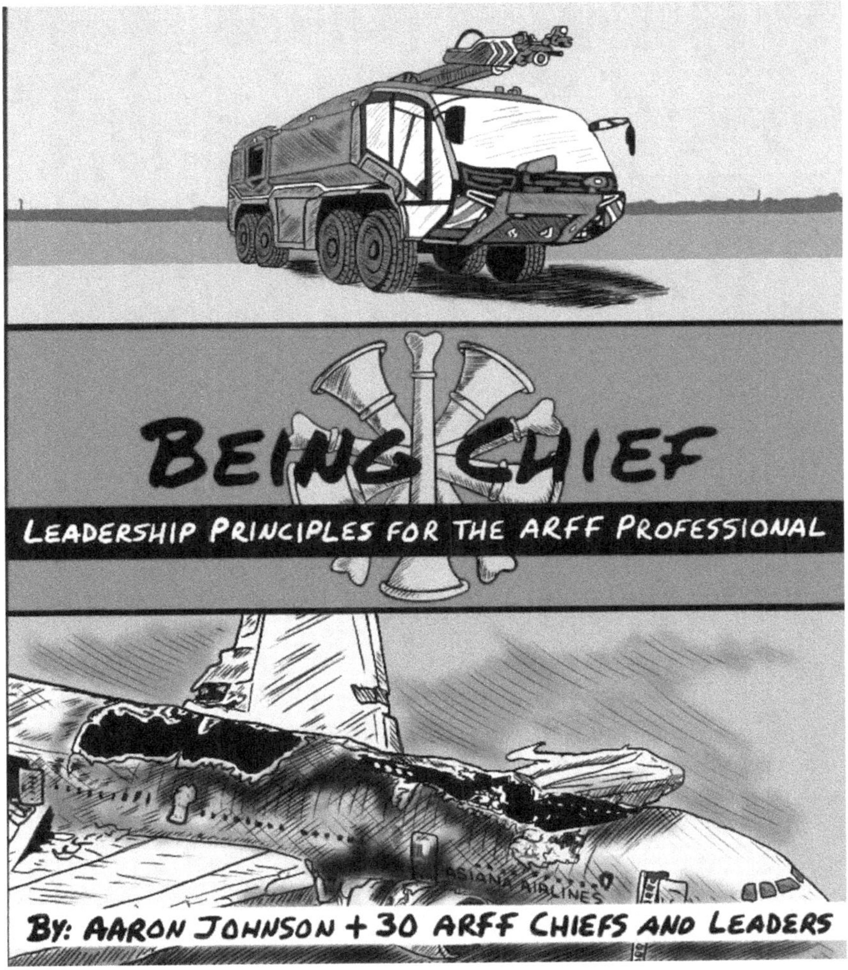

Being Chief
Leadership Principles for the ARFF Professional

By: Aaron Johnson + 30 ARFF Chiefs and Leaders

BEING CHIEF

LEADERSHIP PRINCIPLES FOR THE ARFF PROFESSIONAL

AARON JOHNSON

Copyright © 2021 by Aaron Johnson

All rights reserved.

Cover art and design by Shawn D. Emerson.

Image based on the Rosenbauer Panther 8x8.

No part of this book may be reproduced in any form or by any electronic or mechanical means, including information storage and retrieval systems, without written permission from the author, except for the use of brief quotations in a book review.

www.aaronj.org

CONTENTS

Introduction	ix
Part I LEADERSHIP LESSONS	
1. Leadership Lessons	3
Part II "IN THEIR OWN WORDS…"	
2. Charles Lavene, Fire Chief (Ret.), Norfolk International Airport	13
3. Erik de Laat, Airforce Fire Chief, Royal Netherlands Airforce	17
4. Robert Comeau, Fire Chief (Ret.), Billy Bishop Toronto City Airport	19
5. Lars Andersen, Fire Chief, Fighter Wing Skrydstrup, Royal Danish Air Force	23
6. Stuart Steele, Chief of the Department (Ret.), Trenton Mercer Airport Fire Department	27
7. Mike Evans, Fire Chief (Ret.), Detroit Metro Airport	29
8. John Demyan, ARFF Chief, Lehigh Valley International Airport	33
9. Scott Lanter, A.A.E., Director of Public Safety and Operations, Blue Grass Airport	35
10. Duane Kann, ARFF Regional Sales Manager, Rosenbauer	41
11. Bill Hutfilz, A.M.F. (Ret.)	45
12. William Major, Fire Chief, Buffalo Niagara International Airport	49
13. Jesse Davis, Chief, Airport Police and Fire Department (Ret.), Ted Stevens Anchorage International Airport	61
14. Terry L. Wooldridge Jr., MPA, EFO, CFO, ACE, Fire Chief, Titusville-Cocoa Airport Authority	67
15. Dale Carnes, Fire Chief, Sacramento County Airport	71

16. Andrew Lipari, Captain, ARFF Training Officer, KCFD — 75
17. Christopher Menge, Captain, Training Officer, Albany Airport Fire Department — 79
18. Alvin Lee, Chief, Airport Emergency Service, Changi Airport Group — 83
19. Chris Thain, Business Development Manager, Fire & Rescue Services, G3 Systems Ltd — 89
20. Thomas Littlepage, Captain, Shift Leader, Evansville Regional Airport Safety Department — 93
21. Peter Moore, Manager, Airport Fire Service, Christchurch International Airport Ltd — 97
22. LtCol. Francois Villard, Owner/General Manager, Air Safety and Security Services — 101
23. Elizabeth Hendel, Deputy Fire Chief (Ret.), Phoenix Fire Department — 107
24. Anthony Dynderski, Fire Chief, Sikorsky Aircraft — 113
25. Danny Pierce, Airport Safety Officer (Ret.) — 117
26. Graeme Day, Fire Service Assurance Manager, Capita Fire and Rescue — 119
27. Peter McMahon, Managing Director, Aviation Rescue Services — 121
28. Rob Mathis, Assistant Fire Chief, Portland Airport Fire and Rescue — 125
29. Paul Looney, Sales Representative, Past Chairman, ARFF Working Group — 129
30. Philip DiMaria, Battalion Chief (Ret.), Miami-Dade Fire Rescue — 133
31. Jack Kreckie, Chief of Operations, ARFF Professional Services, LLC — 137
32. Antonio Gutierrez, A.M.F., P.E.M, Fire Chief, Gerald R. Ford International Airport Fire Department — 143
33. Gary Barthram, Chief Fire Officer, Airport Fire & Rescue Service, London Heathrow Airport — 147
34. Joseph Marino, Chief of Department, Port Authority of New York & New Jersey, Aircraft Rescue and Firefighting — 153
35. Mark Huetter, Battalion Chief, NASA/Kennedy Space Center — 157
36. Allen Ward, Regional Director/Command Fire Chief, Rural/Metro Fire Dept., Inc. — 159

Part III
LEADING ON
37. Hall of Mentors　　　　　　　　　　　165
38. Resources for Leaders　　　　　　　　167

Afterword　　　　　　　　　　　　　　169
Notes　　　　　　　　　　　　　　　　171
Acknowledgments　　　　　　　　　　173
About the Author　　　　　　　　　　175
Also by Aaron Johnson　　　　　　　　177

INTRODUCTION

I can still remember my first ARFF Working Group conference. This conference, dedicated to serving the aircraft rescue and firefighting community, brings together the best in the ARFF industry. My first experience at this event was listening to Chief Dale Carnes share about his experience responding to the Asiana Airlines Flight 214 crash. His first person account of the impact this had on his personal life, professional life, and the life of the department is a message that I will not forget. However, I am sure that many people that sat in the same session have already forgotten his words, also there are many new and young ARFF personnel and aspiring chiefs who will never have the opportunity to benefit from hearing this personal experience. As the leaders and the experienced in our profession age out, we are forever losing their wisdom and shared knowledge.

An area of lack in the fire service is in what is known in the consulting industry as "knowledge management". Knowledge management can be simply defined as, "the process of capturing, distributing, and effectively using knowledge."[1] Knowledge management is taking advantage of what is known to maximize an organization's value, or a department's value to the community.

Of knowledge, there are two types, explicit and implicit. Explicit knowledge is data, facts, and captured documentation. Implicit is the knowledge that exists in the heads of people and is only acquired over time through education and experience. This implicit knowledge becomes codified when it is shared through discussions or documentation.

So how do we do this? How do we collect and keep the knowledge of those who have gone before us, who have the years of experience and education? How do we access and apply this knowledge to improve our department, leadership, and response tactics? The consulting firm McKinsey & Company has four principles they use to accomplish this knowledge management objective. We can apply these to our industry[2].

1. Don't reinvent the wheel. Somebody, somewhere, has most likely experienced the same problem that you are experiencing. They have already done the mental exercise of thinking through the problem, and performed the hard task of creating a solution. For practically any problem there is an abundance of reports, documents, spreadsheets, presentations, or graphs that can assist in the solution implementation. Search out these documents. Additionally, there are people within our organizations who are experts at different things. They have different skills in the fire service - rescue, operations, tactics, prevention, command - and outside of the fire service that may be applicable to your problem. Know your people, and utilize their strengths.

2. Develop a rapid response culture. When I started out in my professional career, I was amazed at how easy it was to "wow" people with my customer service. All I did was return phone calls and emails, and follow-up when I said I would. It is very frustrating to be tasked with solving a problem when the people you need information from for the solution are unresponsive. Implementing something similar to a "twenty-four-hour response policy", where any inquiry, in person, by phone, or email is required to receive a response within twenty-four hours, can quickly decrease the time and work it takes to reach a conclusion.

3. Acquire external knowledge. Search out and use experts outside of your organization. Apply knowledge from other industries to solve unique aircraft rescue and firefighting problems. Maintain documentation on the information you collect and add individuals to your database or spreadsheet of experts to consult.

4. Promote knowledge accumulation. Knowledge management should be promoted from the top ranks all the way down. Incentivize rapid response and the support and development of others within the organization. At the completion of big projects or problems, bring the team together to summarize lessons learned, processes involved, and take aways for other operations. Utilize an AAR (after action review)[3] to identify concerns and compile lessons learned after any incident. When members return from conferences or training sessions, bring them together or put what they learned into a shared document, so the whole department can benefit from their experience.

There are many books on fire service leadership, and many more on leadership in general, however, this is the only book that focuses on leadership within the unique niche industry of the aircraft rescue and firefighting environment.

This book is a best effort to jump start a formal knowledge management process in the ARFF industry. For this book, interviews, surveys, and follow-up conversations were conducted from more than thirty ARFF chiefs and leaders. The reader will find that this book is divided into three parts:

> **Part 1: Leadership Lessons.** Common themes and advice. The top advice for leaders is compiled and shared in this section.

Part 2: "In their own words…". This section includes the full interview and survey responses from these leaders.

Part 3: Leading On. This section identifies mentors that have gone before, the value of mentoring, and provides recommended resources for leadership development.

I

LEADERSHIP LESSONS

1

LEADERSHIP LESSONS

AIRCRAFT RESCUE and firefighting professionals from around the world were interviewed for this book. From these interviews with ARFF officers, chiefs, and industry leaders we can garner the most important personal qualities and characteristics that one needs to successfully rise through the ranks to the highest positions in ARFF. The traits that repeatedly rise to the top are: **Trust**, **Honesty**, **Integrity**, and **Humility**. These are great traits to have, and to strive to build into one's own life and leadership. But, what are the practical implications of these? What does trust, honesty, integrity, and humility look like on a day-to-day basis? How do these qualities make one an effective leader? How does the possession, or omission, of these qualities affect the people in the world around us?

TRUST. Trust is defined as, "firm belief in the reliability, truth, ability, or strength of someone or something[1]". In the fire service trust may mean a variety of things including: keeping confidences, having each others back, doing what is right so the public can rely on the fire department when needed, keeping one's word and following through with stated actions, building up others to be effective in their roles and future positions, and knowing that members will fulfill their assignments on an emergency scene[2].

Trust is a two way street. To be a successful chief officer or leader the leader must both, trust his people and *be* trusted by his people. The practical application of this, how this trust is built and fostered, hinges on two critical factors.

The first of these is that the leader must be with his people, the leader makes his people a priority. To be a great leader you must spend time with your people. Listen to your people. Learn your people. Get to know who they are, what drives them, what their dreams, ambitions, and hobbies are. Know where their talents, strengths, and weaknesses lie. Realize and treat each person as an individual. Understand that there are a multitude of personality types. Each type learns, responds, and communicates differently. It is the leaders responsibility to know the individual and how best to communicate with him to achieve optimal results and prepare both, the individual and the department as a whole, for success.

In his book, *The 360° Leader: Developing Your Influence from Anywhere in the Organization*[3], John Maxwell states the importance of walking slowly through the halls.

> "One of the greatest mistakes leaders make is spending too much time in their offices and not enough time out among the people. Leaders are often agenda driven, task focussed, and action oriented because they like to get things done. They hole up in their offices, rush to meetings, and ignore everyone they pass in the halls along the way. What a mistake! First and foremost, leadership is a people business. If you forget the people, you're undermining your leadership, and you run the risk of having it erode away. Then one day when you think you're leading, you'll turn around and discover that nobody is following and you're only taking a walk.
>
> Relationship building is always the foundation of effective leadership. Leaders who ignore the relational aspect of leadership tend to rely on their position instead. Or they expect competence to do 'all

the talking' for them. True, good leaders are competent, but they are also intentionally connected to the people they lead."

The second key to building trust and being trusted is to be clear in your communication. People trust a leader who demonstrates a focused vision, decisiveness in decision making, and clarity in communicating that vision and those decisions. Brené Brown[4] says that to be unclear in our communication is to be unkind to the people we are communicating with.

> "Clear is kind. Unclear is unkind...most of us avoid clarity because we tell ourselves that we're being kind, when what we're actually doing is being unkind and unfair...Feeding people half-truths or bullshit to make them feel better (which is almost always about making ourselves feel more comfortable) is unkind...Not getting clear with a colleague about your expectations because it feels too hard, yet holding them accountable or blaming them for not delivering is unkind...Talking *about* people rather than *to* them is unkind[5]."

Clear communication comes first by listening. Listen to your people, hear what they have to say, know what their deep needs and concerns are. From the point of listening, when your people feel that they have been heard they are open to hearing from you. Theodore Roosevelt has been attributed with saying, "People don't care how much you know, until they know how much you care." When people know that you care for them, that you have their best interest in mind, that they are a priority - this is the point at which trust is built - then they will want to listen to you and will be on board with the department's vision.

Trust in your leadership abilities and capabilities comes by being direct, decisive, and committed to the decisions that you make. With the department and its people as the driving force of your vision, not a personal agenda, then each decision becomes clear and unambiguous. Is this decision the best for the department and its people? Does this move us forward? When the answer to both of these is a resounding "yes" then decisive and clear commitment to decisions made becomes almost automatic.

"Violence of action" is a military term which denotes full commitment to the battle. It suggests the full force, with no restrictions, nothing held back, of speed, strength, surprise, and aggression to achieve total dominance against an enemy. When your people are clear on your communication then they can implement "violence of action" to the department's vision, mission, and goals.

Sun Tzu, writes in *The Art of War*, "But when the army is restless and distrustful, trouble is sure to come from the other feudal princes. This is simply bringing anarchy into the army, and flinging victory away[6]." Without trust, victory in the station, the department, and on the emergency scene will not be had. Rather, defeat will meet the Chief at every turn. With clarity in communication and having a true knowledge of your people, trust is fostered, in both directions.

HONESTY. One of the best definitions of this term is "not deceptive or fraudulent;genuine[7]". Leaders, at some point or many points, may suffer from what has become known as "imposter syndrome"[8]. Imposter syndrome is that feeling of inadequacy that one may experience, that feeling that you do not really belong in the position, or that you do not deserve the rank or title, and that irrational fear that you will be "found out"[9]. Author and teacher, Seth Godin says, "When we embrace imposter syndrome instead of working to make it disappear, we choose the productive way forward. The imposter is proof that we're innovating, leading, creating."[10]

To demonstrate honesty in your character requires you to be honest with yourself. Be honest with yourself that you may be experiencing this imposter syndrome. But, also, be honest with yourself about your knowledge, skills, and abilities. Evaluate yourself on what you do know, a group of people trust your abilities enough to put you into the position and give you the title and rank of 'officer'. You have the skills, you know what to do. Also, be honest about what you do not know. If you do not know something, say so, then do the work to figure it out, find the person who does know. Be honest with others, about your fears, failures, and successes. Honesty, freedom from deceit and fraud, comes through being transparent and vulnerable to your people and to who you are.

INTEGRITY. The word integrity can be defined as sound moral character, who you are when no one is around, and the state of being whole and undivided[11]. Integrity may seem to sum up all the traits - of trust, honesty, and humility - required for an ARFF officer or leader. Though these things may be true, the practical demonstration and application of integrity can stand alone.

This quality of integrity, "of being whole and undivided" is demonstrated in the department through consistency, commitment, and lack of self-serving. The majority of ARFF officers and leaders cited consistency as a key factor to their success. People of integrity are consistent. They show up every day. They show up everyday and give their all, they show up at 100%. They do not give up, they do not quit when the job gets hard or they get offended. They do not leave and go somewhere else whenever the whim hits, they are not always looking for "greener grass". No, effective ARFF officers and leaders are consistent.

Integrity is demonstrated by commitment to the job, the department, and the people. This commitment is practically applied by a dedication to continual learning and professional development. Learn new skills, network and interact with others in the industry, both, ranks above and below. Learn from and integrate skills from other industries outside of ARFF or the fire service. Stephen Covey says, "Some people say they have twenty years [of experience], when in reality, they only have one year's experience, repeated twenty times[12]." The leader must focus on growth, enrichment, and improvement, or he will find himself at the end of his career having lived the same day over for twenty years.

Integrity puts others' needs and the department's first. There is no room for a self-serving attitude in the ARFF officer or leader. In fact, a self-serving attitude is a sure way to failure, as with this attitude there will be no consistency, no commitment, lack of honesty, and no trust will ever be built with your people. One cannot show up consistently every day, and be fully committed to his department and people if he is constantly looking for the next best thing for himself, or how to make only himself better.

Humility. Terms and phrases commonly used to describe the humble person are freedom from pride and arrogance, acknowledging they do not have it all together, seeking to add value to others, and gratitude for what they have. This is practically implemented in ARFF leadership when officers and leaders are approachable, willing to learn, and share the knowledge they have.

The humble leader is approachable. He does not repel his followers with an air of intimidation or stand-offishness. Humility is demonstrated by the fact that the leader's people can openly and freely come to the leader with their needs, concerns, failures, and wins. They can come because they have no fear of being ridiculed or devalued. The humble leader can listen to his people because he knows, and acknowledges, that he was once where they were. He remembers where he came from.

The humble leader is cognizant of the fact that he does not know everything or have all the answers. In humility he is not afraid to approach others and ask for help, guidance, assistance, and advice. When something gets too big or too much, he freely asks for help, and shares the load. The humble leader is not afraid of being "upstaged" by ranks below or above. He understands that his people come from a broad range of life experiences, and there is much to be learned from collective knowledge. Through obtaining and sharing this collective knowledge, everyone gets better.

These traits and their practical demonstration, have significant overlap. The way we show trust is also indicative of our integrity. Honesty requires a significant amount of humility. These four character traits of, trust, honesty, integrity, and humility can be practically applied in these five actions:

1. **Listen**, to those above and below.
2. **Learn**, from others, from past success and mistakes, and from a commitment to continual education.
3. **Be with your people**, they are the priority.
4. **Remember where you came from**, and how you got here.
5. **Be consistent**, always fully show up.

Rising through the ranks to become a successful ARFF leader can ultimately be summed up in the words of Chief Brunacini, "Be Nice![13]", and in the age-old wisdom of the golden rule, "Do unto others as you would have them do unto you[14]".

II

"IN THEIR OWN WORDS..."

2

CHARLES LAVENE, FIRE CHIEF (RET.), NORFOLK INTERNATIONAL AIRPORT

How did you get to your position? What path did your career follow?

- Member of Eatontown NJ Volunteer Fire Department (April 1979-1986)
- Enlisted in the United States Air Force as Fire Rescue (April 1986 to July 1991)
- Hired as a Firefighter at Norfolk International Airport (August 1991); promoted to Lieutenant (October 1992); promoted to Deputy Fire Chief (October 2005); named Fire Chief in January 2006 until retirement on September 1, 2020.

What is your advice for the newly promoted chief officer?

Work WITH and FOR the Fire Chief. Remember, your job is to carry out the policies and procedures of the Deputy Chief and Fire Chief. That doesn't mean you can't ask questions or ask for clarification on a policy or procedure, but once the procedure becomes policy, whether you had input or not, you are expected to carry it out and to convey that to your shift personnel.

What was the most impactful call/emergency you have been on? Why?

The one response that sticks out is a "wildland" type of fire response. While stationed at Altus Air Force Base I was part of the team that responded to an out of control fire that started in one of hundreds of large cotton bails (each bail was the size of a tractor trailer). With sixty plus mile-per-hour winds the fire spread rapidly and visibility was zero. There was very little radio communications and face to face communication was impossible unless you walked into someone. Some vehicles actually drove off the roads and into fields because visibility was so poor. How nobody got run over is a miracle. That response taught me that there may be times when firefighters MUST be able to think for themselves. Therefore, firefighters MUST be well trained and disciplined because there may be incidents where communications with Officers (either by radio or even face to face) may be impossible.

That was a HUGE lesson I carried with me at Norfolk International Airport. Driver/Operators of ARFF vehicles are usually by themselves. This means that if, and when, the first ARFF vehicle arrives on scene, the Driver/Operator MUST have the proper training and discipline to set up on the aircraft correctly and apply agent effectively.

What actions, behaviors, or thought patterns lead to leadership failure?

Lack of Professionalism. Are you treating personnel with respect? Do you listen to their concerns? Are you (the Fire Chief) projecting a professional attitude and demeanor? Are you abiding by the same standard operating procedures as the personnel (i.e., arriving on time or much earlier for shift, is your uniform clean and professional) the Fire Chief sets the example by appearance, attitude, and demeanor.

What is the top trait or characteristic that you believe every chief officer must possess?

Trust, with honesty a close second. They actually work in tandem. Chief Officers MUST be honest with their personnel. Just because you are honest with your personnel does not mean you are being critical. If there are areas individuals need to work on, it is up to the supervisor/Chief Officer to advise them.

What is a habit or routine that you have and how does it help you persevere?

I always took pride in myself for attending EVERY roll call/shift change no matter what "hot topic" issues were going on. I felt it was important to show personnel that I cared and gave them the opportunity to ask questions about anything that may be occurring on the administration level. It also gave me the opportunity to advise them of any airport administration/human resources policies that may be coming out.

What have been the key factors to your success and why?

I've tried to be respectful and open minded to my supervisors. Also, as a result of my Air Force experience, I had the opportunity to work with people from different parts of the country, both military and civilian. One thing I learned very quickly was that there are a lot of fire department personnel that you can learn from to become a better firefighter and, ultimately, Fire Chief.

What are the most important decisions you make as a leader of your organization?

Am I doing everything I can to help improve the organization? Am I providing a work environment that entices the best people to want to work for me and the organization? In the end, the MOST important decision is, am I, as the Fire Chief, carrying on the policies and procedures of the organization and conveying that to personnel?

How do you ensure your organization and its activities are aligned with your core values?

I do not believe this is something that has to be addressed on a daily basis. As a Fire Chief of a one station department, it doesn't take very long to see if my core values are being compromised. It is up to me, as the Fire Chief, to ensure activities and performance meets my core values.

What are you doing to ensure you continue to grow and develop as a leader?

It is EXTREMELY important to keep an open mind to learning and listening. As a new Fire Chief, I thought I was the ONLY Chief to experience a particular issue. What I found, from my peers, was that no matter what issue you encounter, another Fire Chief has had the same problem. I will forever be in-debted to the Hampton Roads Fire Chief's group for their guidance and mentoring. Hampton Roads of Virginia is made up of a number of cities that include, Virginia Beach, Norfolk, Chesapeake, Hampton, Newport News, Williamsburg, York County, and military bases in the area.

In addition, I can not begin to tell the reader how important being a member of the Aircraft Rescue Firefighting Working Group has been to my development as an aircraft rescue firefighter and Fire Chief. (That could be a book in itself!)

3

ERIK DE LAAT, AIRFORCE FIRE CHIEF, ROYAL NETHERLANDS AIRFORCE

How did you get to your position? What path did your career follow?

I started in 1989 as a Crew Commander on AB Volkel. In 1994 I became the Incident Commander. In 2000 I went to AB Twenthe as the Fire Chief. In 2004 I became chief of our fire and rescue training center. Finally, I became the Airforce Fire Chief in 2012.

What is your advice for the newly promoted chief officer?

Stay close to yourselves. Don't judge too fast. Listen.

What actions, behaviors, or thought patterns lead to leadership failure?

Acting to fast. Not doing what you said you would do. Telling lies.

What is the top trait or characteristic that you believe every chief officer must possess?

Stay humble and treat people as you would like to be treated.

. . .

What is a habit or routine that you have and how does it help you persevere?

I often manage on the floor. Don't create large distances. Trust is a key issue.

What have been the key factors to your success and why?

Honesty and being open.

What are the most important decisions you make as a leader of your organization?

In the last re-organization I had to make decisions about reductions and staffing.

How do you ensure your organization and its activities are aligned with your core values?

Preparation, preparation, and last but not least, preparation.

What are you doing to ensure you continue to grow and develop as a leader?

Talking with other managers.

4

ROBERT COMEAU, FIRE CHIEF (RET.), BILLY BISHOP TORONTO CITY AIRPORT

How did you get to your position? What path did your career follow?

I joined the military in 1988 as a firefighter. I was promoted through the ranks and retired as a Warrant Officer in 2010. I deployed to Haiti, for the Minuha mission, CFS Alert and some Rimpac while I was on the west coast. When I retired I joined as a Fire Prevention Officer for the city of Quinte West. In 2015 I took the position in Billy Bishop Toronto City Airport as a Fire Chief and Emergency Planner. I retired in December 2018, and my wife and I have since been enjoying life on our waterfront property.

What is your advice for the newly promoted chief officer?

Be transparent, dependent on your background and experience level. Meet with your team, interview them, ask questions to show you care. Be involved and work with them to see how they operate. Open yourself in a professional way and take notes. Take your predecessor's advice with a grain of salt. Do your own investigation on the issues at the department before commenting. Do not work more than what you are paid for. Share your work load with other officers and fire department personnel. This will give them a chance to learn your position. Use a personal and work phone. Read the SOG's/SOP's. Meet regularly with your management, mutual aid officer, police, EMS and city council. Create an open minded relationship and train together. Always be prepared to be fired at anytime, always have a backup plan if this should happen.

What was the most impactful call/emergency you have been on? Why?

The call was a drowning person near our airport. I was not the first person on-scene but took command after a proper transfer of command. Some basic firefighting issues were observed. By jumping into the cold water without PPE, we were not following our SOG's. This person was already dead and we took unnecessary risk too fast, instead of reacting according to our response procedure. To minimize post traumatic stress, due to the deterioration of the human body in the water, I had a professional come to our department to speak to us.

What actions, behaviors, or thought patterns lead to leadership failure?

Leadership failures are caused by lack of expertise and bad communication skills. Despite this you need to make mistakes and learn from those mistakes.

What is the top trait or characteristic that you believe every chief officer must possess?

Good leadership skills, and a positive attitude. Give constructive comments and be firm in your decisions (despite the fact that you will not please everyone).

. . .

What is a habit or routine that you have and how does it help you persevere?

Be open but not too open, meet your team, from managers to workers. Communicate with them and reward them accordingly.

What have been the key factors to your success and why?

Integrity toward my people and being a good communicator. Being involved, and always helping. This helping created a bond among my peers.

What are the most important decisions you make as a leader of your organization?

The right ones! Do not regret your decision making. If not sure, ask your trusted peers for advice.

How do you ensure your organization and its activities are aligned with your core values?

You follow the rules of engagement and the protocol established by your department. Lots of reading.

What are you doing to ensure you continue to grow and develop as a leader?

Meet people, participate in conferences and do not be afraid to learn new ideas. Learn from other's mistakes, failures, or successes.

5

LARS ANDERSEN, FIRE CHIEF, FIGHTER WING SKRYDSTRUP, ROYAL DANISH AIR FORCE

How did you get to your position? What path did your career follow?

I started as a firefighter after sixteen years in EMS as a military paramedic. As a firefighter trainee I did a good job. After only half a year I was offered the chance to be crew chief. During the NCO-school I was categorized as a student with special skills for leadership. After I returned to the fire department I was offered the position as incident commander/training officer and second in command. I accepted, and because of that I was sent back to school to earn a diploma in leadership. I was promoted to fire chief in 2014. Afterward, I was granted a masters degree in international business communication at The Southern Danish University.

What is your advice for the newly promoted chief officer?

Remember that you are in a new role. Never be afraid to ask older colleagues and never let others be afraid to ask you.

. . .

What was the most impactful call/emergency you have been on? Why?

I have been on multiple calls where I had the dilemma of present danger from hot ammunition and the desire to save air crews and materiel. In those situations everything is on the line. It calls for decisions that you can live with afterward if things go wrong. Up till now I have made the right decisions.

What actions, behaviors, or thought patterns lead to leadership failure?

Not following the leadership pipeline is a sure way to failure. You disappoint a lot of people that way.

You have to believe in yourself. You can't expect others to do it if you don't do it yourself. Talk the talk and walk the walk. Be a man of your word.

What is the top trait or characteristic that you believe every chief officer must possess?

Every chief must be aware of something that we call "leadership pipeline". Meaning that, at every level in the organization you have to take care of your own job. It is very important for firefighters that have been promoted. When your level changes, your tasks change. If you are promoted from firefighter to crew chief, you must avoid taking the work out of the hands of the firefighters, especially during incidents. It is very easy to drop back to your comfort zone when the heat is on, but it is a huge and dangerous mistake. The same goes when you are promoted from crew chief to incident commander. Even I have to avoid regressing to a lower level and interfering in something that should be taken care of by others at that level.

What is a habit or routine that you have and how does it help you persevere?

I intend to be present at as many shift changes as possible. I try to be in contact with as many of my employees as possible every day. I also try to be as open as possible. Of course I can't let people doubt that I am in charge, but they have to know that I always have their back. Troubles are kept in-house as long as possible. We are a family. We support and respect each other.

What are the most important decisions you make as a leader of your organization?

I let all people contribute on a daily basis. I give them as much freedom as possible, but let them be aware that when the bell rings there can only be one boss. That is situational leadership.

How do you ensure your organization and its activities are aligned with your core values?

I keep a good and open dialogue with my firefighters and my incident commanders. Everybody is heard, but at the end of the day they all know that I am in charge. I always follow the orders from the high command, but I always let my superiors know what I mean. My feedback is as important for them as their feedback is for me.

What are you doing to ensure you continue to grow and develop as a leader?

I interact with my network and I am always looking for improvements. I know that I have to change at the same pace as the surrounding society. Otherwise, I will lose track with my employees.

6

STUART STEELE, CHIEF OF THE DEPARTMENT (RET.), TRENTON MERCER AIRPORT FIRE DEPARTMENT

How did you get to your position? What path did your career follow?

I started in ARFF with the federal government and through downsizing I ended up at my current location.

What is your advice for the newly promoted chief officer?

Listen, you got to where you are for a reason. Do not veer from your personal professional model....it is what got you here.

What actions, behaviors, or thought patterns lead to leadership failure?

Generational differences, inability of the chief officer to adequately deal with a multitude of personalities.

What is the top trait or characteristic that you believe every chief officer must possess?

Just be nice....(Ret. Chief A. Brunacini)

. . .

What is a habit or routine that you have and how does it help you persevere?

Jumping in and helping out, from training to cleaning the bathrooms. Don't ask them to do something you think you are too good to do.

What have been the key factors to your success and why?

Ability to communicate and interpersonal skills taught to me by family at a young age.

What are the most important decisions you make as a leader of your organization?

Safety, money will always play a role in safety, this is the dirty secret in the fire service. Don't allow your department to suffer for a decision someone at a desk, counting beans, makes. Stand up for safety.

How do you ensure your organization and its activities are aligned with your core values?

Communication, daily communication, personal communication. If one of the department members is having a bad day just try and help. Even if you can't help, they know that you are willing and sometimes that is the level of caring that makes a difference.

What are you doing to ensure you continue to grow and develop as a leader?

Stay healthy and pursue continuing education, know what they know. Don't become the dinosaur that they expect you to become, keep it fresh.

7

MIKE EVANS, FIRE CHIEF (RET.), DETROIT METRO AIRPORT

How did you get to your position? What path did your career follow?

Worked through the ranks - Firefighter, Sergeant, Lieutenant in Fire Prevention, Lieutenant-Platoon, Fire Marshal, Deputy Chief, Chief.

What is your advice for the newly promoted chief officer?

Take it slow. Get to understand the job. Network with other Chief Officers in other departments. Never, never complain down and never engage in disputes in front of an audience. Take it behind closed doors. Keep your department informed and be transparent.

What was the most impactful call/emergency you have been on? Why?

Two years into the job we had a plane crash during landing in which ten passengers/crew died. A few months later we (same crew) were on duty when Flight 255 crashed on take off, killing 156. These impacted all of us for our entire careers.

What actions, behaviors, or thought patterns lead to leadership failure?

Self serving decisions that have a negative impact on the department (decisions made, "just because you can"). Not knowing your job. As Chief you must understand FAR Part 139. You cannot rely on anyone else in your department to make sure you are compliant with the regulations. If you don't own it you will fail.

What is the top trait or characteristic that you believe every chief officer must possess?

Loyalty, transparency, and honesty.

What is a habit or routine that you have and how does it help you persevere?

Faith. Without it the weight of the job will destroy you. Faith gives you compassion and a clear vision on decisions you are making.

What have been the key factors to your success and why?

Always knowing the job and understanding data. Able to make a solid case for why your department needs what you are asking for. Understanding this is not about you but it is about the department and the men and women you serve. Another is, you must be able to make a decision and own it.

What are the most important decisions you make as a leader of your organization?

Hiring the right individuals that will fit into the department. This is a thirty year decision. Also discipline, this has such a big impact on not only the individual but the department.

. . .

How do you ensure your organization and its activities are aligned with your core values?

We are an accredited department so this keeps us on track for all aspects of daily activities.

What are you doing to ensure you continue to grow and develop as a leader?

Well I'm retiring in 101 days. I'm working now to see that my replacement will have everything he needs to be successful.

8

JOHN DEMYAN, ARFF CHIEF, LEHIGH VALLEY INTERNATIONAL AIRPORT

How did you get to your position? What path did your career follow?

I started as a firefighter and worked my way up to Lieutenant, Captain, and now ARFF Chief.

What is your advice for the newly promoted chief officer?

Prepare for change, it will be a challenge, don't give up.

What was the most impactful call/emergency you have been on? Why?

B-757 diversion smoke on board. Limited man power and mutual aid and not being familiar with the aircraft.

What actions, behaviors, or thought patterns lead to leadership failure?

Subordinates expecting you to work for them. Not being there for the job.

. . .

What is the top trait or characteristic that you believe every chief officer must possess?
Patience.

What is a habit or routine that you have and how does it help you persevere?
Research before reacting.

What have been the key factors to your success and why?
Taking the time to think through every response. Know the answer before responding.

What are the most important decisions you make as a leader of your organization?
Keeping everyone safe.

How do you ensure your organization and its activities are aligned with your core values?
By evaluations.

What are you doing to ensure you continue to grow and develop as a leader?
Continue to learn from others in the field.

SCOTT LANTER, A.A.E., DIRECTOR OF PUBLIC SAFETY AND OPERATIONS, BLUE GRASS AIRPORT

COMAIR FLIGHT 5191 RESPONSE

How did you get to your position? What path did your career follow?

I started as a volunteer firefighter/EMT in my hometown in 1983. Hired on with Blue Grass Airport as a Public Safety Officer (combined police, ARFF EMS department) in 1987. Promoted to Sergeant in 1990, Captain in 1996, Chief in 2001 and Director in 2009. Completed a Bachelors of Science in Fire and Safety Engineering in 1994 and a Masters of Science in Criminal Justice in 2008. Became an Accredited Airport Executive (AAE) in 2010 and hold several IFSAC and EMS endorsements and certifications.

What is your advice for the newly promoted chief officer?

Be the leader you would follow and never, ever forget how to be a firefighter. Across my career I looked at it this way, every promotion that I earned could have been taken away for any number of reasons. The craft of firefighting and the rank of firefighter? No human will ever take those away from me.

. . .

What was the most impactful call/emergency you have been on? Why?

The crash of Comair Flight 5191 at Blue Grass Airport on August 27, 2006.

As to the why...

As awful as that day was, I was right where I was supposed to be at that point in time. I had spent years preparing myself, my people and the airport for the possibility of a plane crash and when it happened, we all did the job pretty much exactly how we had trained. During the response I was honored to lead my people, render the scene safe and start the process of caring for the victims' families by recovering the victims, slowly, methodically and prayerfully. Afterwards, I was blessed to work with the family members in building a memorial to honor those lost.

This August will be the fifteen year anniversary and not a day goes by that I do not think about what we all did that day.

What actions, behaviors, or thought patterns lead to leadership failure?

Pride, talking too much and listening too little, close mindedness, forgetting where you came from, letting your firefighting skills atrophy.

What is the top trait or characteristic that you believe every chief officer must possess?

Being approachable, no matter who needs your time or attention.

What is a habit or routine that you have and how does it help you persevere?

I am an avid reader of many topics. Reading has always been a solace.

. . .

What have been the key factors to your success and why?

Always being prepared for the next opportunity, never forgetting where I came from, and always treating the job with the respect and dedication it deserves.

What are the most important decisions you make as a leader of your organization?

Taking care of my people and their families. Making wise decisions about training, equipment and policy. Developing the department's budget with input from all levels within the department.

How do you ensure your organization and its activities are aligned with your core values?

My number one goal is to listen to people, especially when they are not talking. Second, to never forget that every decision I make, has real life consequences (both good and bad) for the people on the front lines. Third, to remember that I do not know it all. Fourth, do not be embarrassed to ask for help from your people. Fifth, if you can only do one from the list above, make it item number one.

What are you doing to ensure you continue to grow and develop as a leader?

Staying engaged with emerging trends, concepts, and equipment through any number of industry trade groups. Working as a Voting Member of the NFPA ARFF Technical Committee and by working on several industry committees and groups.

* * *

On August 27, 2006, about 0606:35 eastern daylight time, Comair flight 5191, a Bombardier CL-600-2B19, N431CA, crashed during takeoff from Blue Grass Airport, Lexington, Kentucky. The flight crew was instructed to take off from runway 22 but instead lined up the airplane on runway 26 and began the takeoff roll. The airplane ran off the end of the runway and impacted the airport perimeter fence, trees, and terrain. The captain, flight attendant, and 47 passengers were killed, and the first officer received serious injuries. The airplane was destroyed by impact forces and postcrash fire. The flight was operating under the provisions of 14 *Code of Federal Regulations* Part 121 and was en-route to Hartsfield-Jackson Atlanta International Airport, Atlanta, Georgia. Night visual meteorological conditions prevailed at the time of the accident.

The National Transportation Safety Board determines that the probable cause of this accident was the flight crew member's failure to use available cues and aids to identify the airplane's location on the airport surface during taxi and their failure to cross-check and verify that the airplane was on the correct runway before takeoff. Contributing to the accident were the flight crew's nonpertinent conversation during taxi, which resulted in a loss of positional awareness, and the Federal Aviation Administration's (FAA) failure to require that all runway crossings be authorized only by specific air traffic control (ATC) clearances.[1]

— NATIONAL TRANSPORTATION SAFETY BOARD
ACCIDENT REPORT

What lessons did your department learn?

The biggest take-a-way is that in past emergency response preparation drills, we had always planned for a plane full of seriously injured people. With 5191, everyone died upon impact save one, the co-pilot, who my guys rescued. This took a toll mentally, as my folks had no outlet for all the preparation and get it done "energy" when they arrived on scene. Get there, get there, get there to help and then BAM! There is no one left to save.

. . .

What changes were made?

Now, in all trainings, we still teach and push aggressive response, suppression, and rescue operations. However, we also temper it with what we learned in 5191. We have also worked with our mental health providers to give them an appreciation for what 5191 did to our people from this perspective.

Had your training and experience to this point adequately prepared you for this incident? How, why, or why not?

Yes, definitely. I could not be more proud of the people I worked with then and now. I had only been Chief for five years when 5191 happened. Still, in that amount of time, we went through and absorbed the changes caused by 9/11, responded to an air ambulance crash in 2002 (three saves, one DOA, one of the rescues was an actual fire fighter who worked part-time for the air ambulance company). The areas that I continually challenged my people in, then and now, are as follows:

1. Know yourself first. Leverage your strengths and minimize your weaknesses as much as possible.
2. Next, know your crew and as a crew, leverage/minimize strengths/weaknesses.
3. Third, innately know your equipment, (from ARRF truck to HAZMAT suit) as well as aircraft and airfield.
4. Lastly, I ask them to remember, something that I was taught as a young volunteer: If you're not helping a scene get better, then you're wasting every professional dime that has ever been spent on you.

What impact has this had on your career and leadership?

I'm approaching my 38th year as a fire fighter, my 34th year as an ARFF and my 20th year as Chief. Two things have always stayed with me throughout my career. First, the last bullet point above. The second one is this, I always remind myself to never forget that <u>I earned the rank of fire fighter.</u> As people promote up, I've seen firsthand that they are more exposed to organizational "fickle winds" which may lock someone into a rank for a career OR take the rank away entirely. To be clear, short of me doing something extremely stupid, <u>no human being will ever take away my being a fire fighter.</u> I earned it and God willing I will be the one to let it go, hopefully when I retire. Watching an officer forget what it means to be a firefighter, shy away from getting dirty, or forgetting how to use the basic tools of our profession is anathema to me.

What advice would you give to prepare someone for responding to an incident of this magnitude?

Prepare for it continually, pray that it never happens, and if it does happen remember that good, bad or indifferent you are exactly where you were meant to be the moment you raised your hand and took your oath.

10

DUANE KANN, ARFF REGIONAL SALES MANAGER, ROSENBAUER

How did you get to your position? What path did your career follow?

I started with a tour in the U.S. Air Force as a Fire Prevention Specialist, then transitioned into the civilian sector as a firefighter at Orlando International Airport. I worked my way through the ranks to become Deputy Chief at thirty-five years old and then Fire Chief at forty. I spent seven years as Chief before retiring to work for Rosenbauer.

What is your advice for the newly promoted chief officer?

Be humble and do not try to make your mark in the first few months. Take time and identify a few target issues and come up with a plan for the personnel to help address those items. You will want to do them yourself for the credit and respect, but you will get a much better and longer term result by getting the team to help and take some ownership in the projects.

. . .

What was the most impactful call/emergency you have been on? Why?

I responded to one of the worst fatality car accidents in my twenty-nine year career on my first day at Patrick AFB. I saw how well our team worked together as I stood by on the nozzle. As I observed extrication, CPR inside the car, blankets over the fatality, I was wondering to myself if I could handle this job. When we got back to the station my Lieutenant on the engine pulled me aside and said not every day is like this, and when we are able to save the victims the feelings are much different. Not sure how he knew he needed to say something, but I am glad he did. I tried to continue that on for any new firefighter that had a bad call.

Another time I was riding out of grade as the shift commander, and we had a potential hostage situation onboard a full aircraft sitting on the taxiway. Everyone on the call, from the police to the airline station manager, was new or working in a higher classification. We were all deer in the headlights, but we worked well together for several hours. SWAT made entry and apprehended the suspect who ended up being unarmed.

What actions, behaviors, or thought patterns lead to leadership failure?

Lack of organizational structure and failure to delegate with appropriate oversight.

What is the top trait or characteristic that you believe every chief officer must possess?

The ability to put yourself in the other person's shoes. My shift commanders have told me that, at times when I was not available, when they had to make a decision the first thing they thought about was WWDD (What Would Duane Do?). At the end of the day everyone has to step up and make their own decisions, but this way you are not making it in a vacuum.

What is a habit or routine that you have and how does it help you persevere?

My daily bike ride allows me to think about the topic of the day while also releasing endorphins to help me feel good. An added benefit is firefighters will respect your decisions more when they feel like you can still do their job. I heard this first hand on a regular basis when I was on the floor as a firefighter.

What have been the key factors to your success and why?

A lot of luck to start with. Years in the industry and learning as much as I could along the way. You have to be educated, but you also have to carry a lot of common sense. Some people just do not have the common sense or people skills to lead in a way that others will want to follow.

What are the most important decisions you make as a leader of your organization?

Every decision is important because it impacts someone, somehow. It was sometimes the smaller insignificant decisions I made that had the most traction on the floor. Not all were good, but some that were not good were necessary. The key to a decision is not just about making people happy, it is about what is right and understanding how that answer will impact the people. Then the decision can be massaged, or other decisions can be included to lessen the negative impacts.

How do you ensure your organization and its activities are aligned with your core values?

A good leader will be clear about what they hold important and will take action (good or bad) when those values are displayed or violated. This will take years if the organization is not already aligned, but it is possible if values are shared on a regular basis. My Deputy Chief made a small card for each person to carry with some simple values and decision making practices. It was met with mixed impressions, but it got a lot of chatter and people knew what he/we expected.

What are you doing to ensure you continue to grow and develop as a leader?

I assess my decisions and actions every day. I do a lot of self evaluation and hope to learn from my mistakes. I sometimes get to go back and ask for a do-over. Most people appreciate knowing you thought about what happened and realized you were wrong. So many people will not admit when they are wrong because they see it as a sign of weakness, but it is actually one of the strongest things a leader can do.

BILL HUTFILZ, A.M.F. (RET.)

How did you get to your position? What path did your career follow?

I retired as the Aircraft Rescue Firefighting Training Officer, Clark County Fire Department where I oversaw all ARFF Training and Operations for the Clark County Department of Aviation, which owns and operates McCarran International Airport and four general aviation airports. I am also a Retired USAF Firefighter (Master Sgt.), Airport Master Firefighter (A.M.F.) and ARFF Working Group Legend. I have more than forty-six years in the fire service.

What is your advice for the newly promoted chief officer?

I always learned from my firefighters. Listen to them, remember you can't ask them to do something if you have never done it yourself.

What was the most impactful call/emergency you have been on? Why?

I have one that sticks out to this day. We had a container fall on a guy killing him. This was pretty stressful for most of my guys. It was a learning experience and training experience. This also allowed us to talk with our Fire Chief and get better equipment for us to do our job of protecting our customers.

What actions, behaviors, or thought patterns lead to leadership failure?
You need to be a people person, don't demean your people. You don't have to be their friend but treat them fair and equal.

What is the top trait or characteristic that you believe every chief officer must possess?
Trust your personnel. They make you, you don't make them.

What is a habit or routine that you have and how does it help you persevere?
Alway be on time or early. Being punctual has always been something that lets people know you want to be engaged.

What have been the key factors to your success and why?
My key factor was knowing my job, and staying up on the latest and greatest innovations in the ARFF world. I was never afraid to reach out and ask for help if I didn't know the answer.

What are the most important decisions you make as a leader of your organization?
A leader leads, but if not done properly the people will do what they want. Give them some ownership in day to day decisions. Listen to what they say, good and bad.

. . .

What are you doing to ensure you continue to grow and develop as a leader?

I'm retired, but still try to stay engaged by helping new leaders who are coming up in the ARFF community.

12

WILLIAM MAJOR, FIRE CHIEF, BUFFALO NIAGARA INTERNATIONAL AIRPORT

CONTINENTAL CONNECTION FLIGHT 3407 RESPONSE

How did you get to your position? What path did your career follow?

My path is one that is probably atypical. I started my firefighting career in a small rural fire department as a volunteer. Like most departments the majority of our calls were EMS. I started on the EMS track which led me into a career in commercial EMS. I remained an active firefighter in the volunteer world. I received most of my management training as I progressed through the ranks in the commercial EMS world. I also gained valuable training working in the regional EMS agencies. I continued my firefighting career and worked my way from EMS Captain to an Assistant Chief with my volunteer company. As fate would have it, my EMS management career came to an end. I started working as a paramedic for the City of Batavia Fire Department. When the opportunity to transfer to the fire side of the department presented itself, I graciously accepted. Then an opportunity to become an Airport Fire Fighter at Buffalo Niagara International Airport Fire Department was presented. I accepted that position in 2008. In 2009, our Chief retired and one of the captains was appointed Chief. The same year two other captains retired.

We are a department of forty-one total. I was selected to fill one of the vacant captain positions. I was placed in charge of the 4th platoon. A platoon of seasoned employees who, to say the least, were not happy with the decision. Over time, and being consistent with my leadership, things improved. During my time as Captain, I filled in as the Acting Chief twice as the Chiefs retired and replacements were sought. After Chief McDonald retired, I made the decision to apply for the Chief's position. In June 2017 I was selected to be the next chief. I am the seventh Chief of our department.

What is your advice for the newly promoted chief officer?

You need to master situational awareness. Not only for incidents, but also for administrative needs. Incidents will run themselves, but with quality leadership they run a lot smoother. As you move up the leadership track, your "incidents" will not be the same as running an EMS run or fire call. They will require training, life experience, and situational awareness to another level. You need to be consistent with your decisions. You need to be fair and able to admit when you are wrong. You need to be able to check your ego at the door most days, and realize there may be more than one way to handle an incident. You need to able to make a decision, even if turns out to be a wrong one. Gather as much information as you need, never stop learning. Remember everyone has a boss.

What was the most impactful call/emergency you have been on? Why?

Over my thirty plus years of being part of the Fire/EMS services there have been many calls that I have been impacted by. It is difficult to select just one. From an ARFF perspective, I was working the night of Flight 3407. I was a firefighter assigned to our mutual aid ARFF truck. The plane crash was in my community where I had lived for twelve years leading up to that night. On that night our Captain was off, and one of the other Captains was working for him. I had a year on the job as an Airport Firefighter. I responded with a senior firefighter, who had not seen much action during his time there. Responding to an incident in the town you now reside in was intense. The days and nights dragged on, as a small department we all had many roles. It gave me the opportunity to demonstrate my leadership skills to the leadership at that time. It helps me as the Chief, to push for more training and for everyone to know their different roles. It has helped me to drive the need to expand our circle of needed help. It also helps to remind me of the impact calls have on the responder, their families, and the community.

What actions, behaviors, or thought patterns lead to leadership failure?

Foremost, is not being consistent with decisions. This does not mean, never changing your path if the plan of action is not working. You need to make decisions. Being unwilling to do the little things. You need to be able to follow orders, even if you disagree, as long as life safety is not an issue. You need to have a plan B, C, D, E, F, etc. in case plan A does not work. Not taking the time to learn about your staff, co-workers, customers, mutual aid partners. You will fail if you think you know everything. No one person has all the answers. Trust your gut. Know that there are many ways to tackle a challenge. Learn from past challenges.

What is the top trait or characteristic that you believe every chief officer must possess?

The ability to listen and not formulate a response while "listening". Take in the information being presented, and process it. Treat everyone with the same dignity and caring you would expect for yourself.

. . .

What is a habit or routine that you have and how does it help you persevere?

I make it a point to take a few hours for myself during the week. Typically, on the weekend, I take my camera and head to the woods. While driving out to the woods, I listen to inspirational music or drive in silence. While I am in the woods looking for that perfect picture I slow down to enjoy nature. I reflect on past activities of the week, the needs of my family, and my needs. Photography and nature resets my brain. This routine came about by recognizing my PTSD and its effects on me.

What have been the key factors to your success and why?

I believe there are a few key factors. Early in my life my parents taught me the value of hard work, the value of not giving up even when you want to. The drive to make things better, even when you feel the world is against you. Push on. Learn that when someone says no, it doesn't have to stay that way forever. Re-evaluate the reason for the no answer and determine if there is another way, do this respectfully.

What are the most important decisions you make as a leader of your organization?

This is an interesting question. Balancing the needs of the department versus the needs of the Authority. Doing what is best for the firefighters, leadership staff of the department, and the Authority. Learning just how one decision can have so many ripple effects, whether with other departments immediately or years later. As Chief, I am involved with union contracts in all aspects, budgets, day-to-day operations to long term operations. My role has expanded from the first day I started to today. I am involved in Uniform Code Compliance for the entire Transportation Authority, not just the airport.

. . .

How do you ensure your organization and its activities are aligned with your core values?

By reviewing them often. Not only with myself, but with many others. One must define what their core values are, with others. I often will ask my staff, family members, other supervision from different groups, if a decision I made was in line with the core values I believe in.

What are you doing to ensure you continue to grow and develop as a leader?

Learning from others. Participating in various trade organizations. Expanding my horizons outside of the fire world. Enhancing my communication skills. Working on my PTSD and sharing the need for other leaders to do the same.

On February 12, 2009, about 2217 eastern standard time, a Colgan Air, Inc., Bombardier DHC-8-400, N200WQ, operating as Continental Connection flight 3407, was on an instrument approach to Buffalo-Niagara International Airport, Buffalo, New York, when it crashed into a residence in Clarence Center, New York, about 5 nautical miles northeast of the airport. The 2 pilots, 2 flight attendants, and 45 passengers aboard the airplane were killed, one person on the ground was killed, and the airplane was destroyed by impact forces and a postcrash fire. The flight was operating under the provisions of 14 *Code of Federal Regulations* Part 121. Night visual meteorological conditions prevailed at the time of the accident.

The National Transportation Safety Board determines that the probable cause of this accident was the captain's inappropriate response to the activation of the stick shaker, which led to an aerodynamic stall from which the airplane did not recover. Contributing to the accident were (1) the flight crew's failure to monitor airspeed in relation to the rising position of the low speed cue, (2) the flight crew's failure to adhere to sterile cockpit procedures, (3) the captain's failure to effectively manage the flight, and (4) Colgan Air's inadequate procedures for airspeed selection and management during approaches in icing conditions.[1]

— NATIONAL TRANSPORTATION SAFETY BOARD
ACCIDENT REPORT

What lessons did your department learn?

We learned a few lessons from this incident.

The first lesson we learned is that it can happen, even when most tell you it cannot. From the first days here at the fire station the older firefighters and many other seasoned airport employees would often say it is not going to happen here. It happens elsewhere. I believe this is due to the infrequency of these types of incidents which lulls people into a false sense of security. We also learned that our airport emergency plan is written for on-airport emergencies and not designed for off-airport emergencies. We were an assisting agency not the home agency which brought some challenges to the operation overall. We did not have the final say on how to manage the incident, and could only make suggestions to the non-aviation world of emergency services. We were able to accomplish what we needed too, but it did require extra work.

We did not have a plan in place to transport the replacement foam out to the crash site, which was roughly seven to eight miles away. At the time, our foam was in five gallon totes. We figured it out and it arrived. But was it as efficient as possible? What was missed because we were thinking about something that we had plenty of time to evaluate before it happened?

We did not have a plan for the mental health piece for anyone other than a firefighter or police officer. The airport did do a mass stress debriefing a couple of days later. Every person or group (terminal, airfield, fire, police, and administration) was brought to the fire house and sat in one group.

Our terminal staff and TSA were not prepared to handle the

emotional piece brought on by the friends and family waiting at the airport. Yes we have a plan where to place the friends and family on paper, but that doesn't cover the emotions displayed. Our clergy portion was not broad enough.

Our union was not prepared to counter claims from the Buffalo Firefighters Union when representing who had what role with the incident. The press release claimed Buffalo Fire had responded initially to the incident and provided the support. The International was contacted. The response was- "We were told they were". They did no fact checking. We left the union shortly after that.

Because we have been through this, it is easy to have a false sense of security. Often you will hear, "we handled 3407". Which we did, however, it wasn't on our property, EMS and the hospitals had such a small role due to the number of fatalities.

What changes were made?

We changed from five gallon totes to 275 gallon totes. We also purchased a trailer and converted it into a foam trailer. The foam trailer has two 1500 gallon monitors and three 275 gallon totes. This trailer can be easily moved on and off property if needed. It does require the assistance of our airfield, but is doable.

We continue to work on the mental health piece. This is something I could go on for hours about. I believe it be one of the biggest items we, as an industry, need to acknowledge and work on. I have worked with our Airfield Superintendent, the Health and Safety Director, and our EAP program to work on a program for those not currently receiving CISD services. It sounds like an easy fix, but has had some hiccups.

We have adjusted or AEP review and full scale exercise to include all the different aspects a crash will bring. Our last two full scale exercises have included having twenty to thirty actors placed at the terminal to act as friends and family. This has been an eye opener for the terminal staff and TSA who, in our plan, handle this piece.

We acquired a Boeing 727 from FedEx to enhance our training for incidents in 2012. We are still using it today. We have our mutual aid departments, law enforcement, and deicers use it. We have used it for full scale exercises, as well. We have asked our clergy to work on expanding their group to cover more religions.

I use the phrase "everyone said 3407 would not happen here, and we know how that worked out" when I receive push back on preparing for an incident that may or may not happen. Just recently with our terminal expansion we were being trained on a piece of equipment used for passengers to pass through to exit the secured area into the unsecured. It does have the ability to secure passengers inside of it. I asked how to unsecure them. The trainer stated, "you will never have to do that because its never happened before". I started responding with the standard "I have a universal key" which he again responded, "you won't need that". I responded, "People said 3407 would never happen here, but it did. So, rather than debate the never going to happen show us how to respond when everything else fails". It took him two minutes to show us what to do. He spent fifteen minutes arguing why he didn't need to.

I have pushed EMS and the hospitals to realistically answer questions dealing with how many patients they can handle at one time. Again I will hear, "we handled 3407", which I then remind them of only one transport. So did we handle it? I will often ask how many ambulances are available. I will be told a number, for example thirty-four. When I ask how many are on calls, the number is twenty-five. How many are dedicated to other contracted municipalities and such? I may hear six. So really how many ambulances do I have available in the first fifteen to twenty minutes? Same concept with the hospitals and trauma room beds and staffing. We know to look at it from a start to fifteen minutes, fifteen minutes to thirty minutes, and keep adding time and readjusting resources. This has helped push the idea that no one can do it all.

Had your training and experience to this point adequately prepared you for this incident? How, why, or why not?

Great question. Did my training as an ARFF firefighter help me? I started working for the NFTA on 1/28/2008. 3407 happened on 2/12/2009. I had been off probation less then a month. My captain was off that night, and the captain on duty I had spent very little time working with. My forty hour Basic ARFF training did help to a point, the on the job training certainly made me familiar with operating the equipment. My training in leadership with past emergency services did help prepare me. At the time of the incident I was an Assistant Fire Chief with Clarence Fire District #1 which is the first due department with Clarence Center Fire Department. Our two departments, along with many of the other fire departments that responded, had trained together in the volunteer fire world. There had been no training between the airport fire department and the other departments who responded that night. It was brought up and various seminars where Clarence Center Fire Chiefs spoke referred to us training together, but that was not the case. Maybe he was referring to the volunteer training.

I believe my training helped me assist the volunteer firefighters to deal with such a large and heartbreaking incident. The plane crashed inside the Village of Clarence Center not far from their fire station. There was a large amount of residents early on, and through the night, until law enforcement could get a handle on securing and moving people back behind the fire line tape. I was able to direct the volunteer firefighters how to handle the cooling down of the scene after we had applied AFFF to knock the bulk of the fire down. Initially, many of them were treating the scene as any other large fire. Many used their 2.5 inch line with 150 psi to attempt to "breakup the piles". The piles were burned torsos of the passengers.

We needed to organize a line of firefighters to resupply the ARFF truck with AFFF. This was accomplished by lining up firefighters and passing the buckets to one another.

Overall my experience as Assistant Fire Chief and General Manager of an EMS agency gave me the insight to remain calm, prioritize, delegate, and analyze the effect of our actions under the Unified Command established by Clarence Center.

The following morning I was sent to the Erie County EOC as the representative. My ICS 300 and 400 training was a tremendous asset for myself. I had a good understanding of how the EOC would operate at this level of an incident. I also learned that not everything we are taught is followed. Up until the NTSB arrived there was a joint effort to have all the agencies work together and communicate the necessary messages. Once the NTSB arrived that all changed, and with good reason. The NTSB remained sterile from our EOC and operated their own stand-alone. They took over releasing information and directing the investigation.

What impact has this had on your career and leadership?
This incident has had a tremendous impact on my career at the NFTA. I believe I would not be the Chief at this very moment had I not been assigned to F4, our mutual aid truck. It provided me an opportunity to demonstrate to my Assistant Chief and Chief, at the time, how I could operate under extreme circumstances. It allowed me the opportunity to demonstrate my problem solving skills, and the ability to make a decision and follow through with it. The airport management also saw those traits. When it became time to be evaluated for a promotion, I believe that was a huge help.

It has changed my leadership style as well. It provided me with confidence, but also a reminder not to become complacent. It demonstrated the need to be able to adjust your staff to meet the needs of the incident. The need to think outside the box. Not every one reacts the same way when the incident is happening, or months after it has happened.

I would also say it negatively impacted me, as well, for some time. I did not take the necessary breaks that I should have. I did heal the mental side of the incident. I still often have dreams of plane crashes in my sleep. It was a few years after the incident I started my treatment for my Complex PTSD. I have learned what most of my triggers are and how to counter them. For the longest time, I used the wrong coping mechanism. It impacted my personal life, like so many others that night that responded. For many it was their "Ground Zero", for me it was another call in the thirty plus years I have been doing Fire and EMS. I had to learn how to understand and lead those who live and die by that night. It's a great tool in our tool box for handling large scale horrific incidents but it is not the only one. I have seen too many use this incident and not recognize the things that didn't occur that normally do. An example being a large debris field, and multiple levels of injuries.

What advice would you give to prepare someone for responding to an incident of this magnitude?

My first piece of advice is take two deep breaths when dispatched and before you roll out the door. You need to have a clear head when responding.

I drive situational awareness all the time to my staff. In many of the conversations I have with my staff and others at the airport is I say to stop thinking it will never happen. Ask yourself often if this, fill in the blank, happened right now what would I do? I preach know your district, your capabilities, those you lead, those you depend on. Look around every day and see what is there. Plan for all phases of the incident. Be prepared for this incident to live for days. Many focus on the crash itself, our part with life safety should be wrapped up within a few hours. Either all the wounded are transported and treated, or they are deceased. As I wrote earlier, know the true number of mutual aid assets you have. Drill down on them and confirm. Train on all phases of the incident and build your plans.

Unified Command is a huge component. Realize your role as the incident plays out and changes. There are four phases to every incident:

1. Making preparations before it happens to prepare for when it does.
2. Responding to the life safety component. The fire department will be the lead.
3. Investigation of the scene and incident. Law Enforcement,

FBI, FAA, NTSB are in now in charge. Not many providers have daily interaction with the FBI, FAA or NTSB. The fire department role is a supportive one.
4. Bringing the area back to where it was prior to the incident. The fire department role is supportive.

When you look at incident we have very little time of being the lead comparatively. But we also are probably going to have the most time on scene.

Think about the human needs that need to be provided. Food, water, lavatories, and places to rest. You will probably require mutual aid to assist throughout.

Mental health, the elephant in the room. This area needs its own chapter. Think about those affected. We do a so- so job with our staff. What about their families? What about the ATC controller(s), the airfield staff, terminal staff, the airline staff, the vendors at the airport, the on-lookers the community and their families? These are very labor intensive goals to meet and they require a lot of pre-planning. You cannot just whip it up and hope it works. My daughters, who were much younger, were impacted and I couldn't see it. All four worried I wouldn't come home when they saw the footage on TV that night. My two youngest live in the community and saw all the outpouring from the community. Be prepared to accept the help they are going to want to give. It was incredible to see and be a part of. Writing this brings backs moments of joy and sadness. Before getting the help I needed, I would have called myself weak and told myself to shut up and move on. It's part of the job. I have learned now to grieve the loss, allow a moment of tears if I want for those lost. Come to grips with the fact that I didn't cause the incident and that I did the best I could. This causes me to push myself to improve my skills because you don't know when the next incident is coming. We do know it's on its way though.

JESSE DAVIS, CHIEF, AIRPORT POLICE AND FIRE DEPARTMENT (RET.), TED STEVENS ANCHORAGE INTERNATIONAL AIRPORT

How did you get to your position? What path did your career follow?

Following a career in the Air Force, I joined the airport police and fire department. I worked twelve years as a Firefighter/Patrol Officer. I served as an EMS instructor and field trainer for new recruits. I became the department Infection Control Officer, then became head of the Training Department from 2008-2011. In November of 2011, I was promoted to the Chief of the Department, where I served from 2011-2021.

What is your advice for the newly promoted chief officer?

Too often, new chief officers try too hard to hang on to the excitement of their "former" firefighting career. Accept that you are starting a completely new and different career. You are a leader now. Hopefully, you have the technical skills and firefighting education; now you need to immerse yourself in reading and learning about leadership and supervision. Entire libraries and bookstores are filled with books on this topic. Become a student of leadership; read online articles, and attend classes. Never stop learning or seeking to expand your knowledge of how other departments in your city or other airports operate.

What was the most impactful call/emergency you have been on? Why?

A cardiac life save. All the training came together in one moment. Everyone worked effortlessly together, each taking on a role wherever we were needed. This instilled in me the need to guarantee that level of performance continued in every level of the organization, from our treatment towards the public, to emergency timed response drills, to the readiness of our equipment.

What actions, behaviors, or thought patterns lead to leadership failure?

You cannot hide your behavior, people will notice. One of the earliest lessons I learned in school and my military career is the concept of "The Appearance of Evil". Right and wrong do not matter, perception kills. You don't get a chance to explain yourself. If it looks wrong, it is wrong. If you favor a friend over another coworker, even if you are 100% in the right, you will fail. If you are self-serving and pumping yourself up at the expense of your firefighters, you will fail. You must be completely dedicated to the department and question your every decision to ensure it is the best one, at the time, for the department.

What is the top trait or characteristic that you believe every chief officer must possess?

Trust. The department members must know that you are there for them. They can accept that you make mistakes, but they will follow only if you demonstrate selfless service and that your intent and goal is always with them and the department in mind. They may not like your decisions, but they will respect that you didn't make them for personal reasons, or that you show favoritism. Get out of the office regularly and chat with the firefighters, dispatchers, and administrative staff. Get to know them and be sincere about their goals and interests, both on and off the job.

What is a habit or routine that you have and how does it help you persevere?

As a firefighter, you had your good days and bad. As the Chief, you are not allowed to have a bad day. Even in the foulest of moods, your job is to lead, to inspire, and to motivate. You cannot do that if you are pouting, or mad at the world. You can be firm and demanding, but your mood will rub off on others. Complaints go up, not down. Don't share the frustrations that you are having with your boss with the firefighters or your office staff. Occasionally, step back and look at the positives and the progress that you have been a part of. Few will recognize you for those. You have to do it for yourself, and then keep pushing hard for the next positive impact.

What have been the key factors to your success and why?

Having been the Chief for ten years, I am still a learner. Success is not a goal, and must always be chased. When you reach one intermediate goal, you must develop another. There are very few days when my calendar is not full, or I'm dealing with some small personnel crisis, or deadline. Some of the best days are when I have a moment to look forward, over the horizon, and develop the strategy for the next month or the next year. Always look for the next project to reset the goal post. People think that the Chief just sits in the big office. Sitting has never been so exhausting!

What are the most important decisions you make as a leader of your organization?

Goal setting for the department. We must have a plan for the week or the month. Even if that gets interrupted (and it will), we cannot just get up in the morning with the intent to just "come to work". I depend on my lieutenants and captains to be my advisors and I include them in decisions, mentoring them to take an interest in command and executive decisions above their levels. I assign them additional duties or major investment projects that will have generational impacts for the department. I want them to take ownership of the department.

How do you ensure your organization and its activities are aligned with your core values?

My vision as the Chief was to be the "best department in the state - the best equipped, the best trained, and the most respected". I repeated that until my captains and lieutenants adopted that vision and made it their vision. We developed a mission statement and displayed it in the department. I explained it to the firefighters and explained the need to help row the boat towards that. I explained that I expected there would be some who would not put their oar in the water, but I do not accept or tolerate someone who merely sticks their oar in the water to slow us down. I needed to demonstrate my sincerity by providing training at all levels with the mantra that, "If we are not dealing with an emergency, then we should be training for that emergency." I needed to send firefighters, not just chiefs, to conferences. I needed to hear from the firefighters what equipment they needed and then get it for them. I emphasized that we are public servants above all else. I began annual ethics training throughout the department and slowly pushed that out to the Battalion Chiefs to take on that role of teaching. I recognized the need to start developing my replacement from day one, and I urged my other chiefs to prepare for their eventual replacement as well.

What are you doing to ensure you continue to grow and develop as a leader?

Accept that I will never stop learning. Seek out leadership education at all levels. I must emphasize that chiefs need leadership training and education as much as firefighters need the technical skills. I began sending chiefs and lieutenants to out-of-state leadership training. I then had them share what they learned with the rest of the staff. I learned alongside my lieutenants and captains. If I read a good leadership book, I shared what I read with others and encouraged them to do the same.

TERRY L. WOOLDRIDGE JR., MPA, EFO, CFO, ACE, FIRE CHIEF, TITUSVILLE-COCOA AIRPORT AUTHORITY

How did you get to your position? What path did your career follow?

Started out in the military (six years Active Duty / fourteen years IMA Reserve), then eleven year's at City of Melbourne Fire Department, and now seven years at Titusville-Cocoa Airport Authority

What is your advice for the newly promoted chief officer?

Adaptive change takes time, be patient. Surround yourself with five people who you fully trust (not yes men/women) and allow them the ability to disagree with you in order to make the organization better.

What actions, behaviors, or thought patterns lead to leadership failure?

Ego, attitude, lack of integrity.

. . .

What is the top trait or characteristic that you believe every chief officer must possess?

The most important trait all chief officers should have is the ability to remain calm and not react in anger when provoked.

What is a habit or routine that you have and how does it help you persevere?

I find on days where I work out early, before starting my day (0500), helps me mentally prepare. This time allows for me to collect my thoughts and reply to emails as I am on the elliptical or treadmill. I can also catch up on national news. Copious amounts of coffee help too!

What have been the key factors to your success and why?

The true key to my success has been the tremendous teams I have had the opportunity to lead. Without them, and their ability to think outside the box and be completely open to change, my ability to manage and ultimately lead our organizations would not have been possible.

What are the most important decisions you make as a leader of your organization?

The most important decision any leader of an organization makes are the ones that place the department in a positive light. Making decisions that are ethical, consistent, and by the book, or that follow protocol or rules/regulations, is vital to success!

How do you ensure your organization and its activities are aligned with your core values?

Our agency is constantly reviewing what we do versus what we say we do (Core Values, Mission Statement, etc.). If something is in question, we generally sit-down and evaluate as a committee how we can realign or discontinue the activity.

DALE CARNES, FIRE CHIEF, SACRAMENTO COUNTY AIRPORT

How did you get to your position? What path did your career follow?

I have thirty-six years in the fire service. From Las Vegas Fire and Rescue to San Francisco Fire Department, where I retired as an Assistant Deputy Chief. After retiring from SFFD, I went to Sacramento Airport Fire as the Fire Chief. During my career I worked as an ARFF Lieutenant, ARFF Shift Captain, and ARFF Chief at SFO and as the Fire Chief at SMF

What is your advice for the newly promoted chief officer?

Don't lie to yourself that it will slow down in a few weeks, months, etc. and then you'll catch your breath. The op tempo you are experiencing is what it is, it's what we do as chief officers. It will slow down after you retire.

What was the most impactful call/emergency you have been on? Why?

The Asiana 214 crash at SFO, July 6th, 2013. It involved the crash of a Boeing 777 with over 300 souls onboard, a working aircraft fire, actual physical rescues from the burning aircraft, and an MCI with over 180 patients. I was not only involved in the response and aftermath, but also was a member of the NTSB investigation team and participated in eight hours of testimony before the NTSB in Washington, D.C.

What actions, behaviors, or thought patterns lead to leadership failure?

Unwillingness to communicate (high priority on listening) with your people, a silo mentality, and a lack of caring for your people.

What is the top trait or characteristic that you believe every chief officer must possess?

A servants heart.

What is a habit or routine that you have and how does it help you persevere?

When I am developing a policy or procedure change, or any major change for our agency, I meet with my junior chiefs, lay out the plan for them and then ask them to shoot holes in the plan and look for unintended outcomes or anything I missed. This habit greatly reduces the number of "do-overs" I have to deal with (always costly for a Fire Chief), and it greatly increases buy-in.

What have been the key factors to your success and why?

My faith, my family, my military experience, and some incredible mentors.

. . .

What are the most important decisions you make as a leader of your organization?

Any decision that will affect the long term future of the organization and the welfare of the personnel.

How do you ensure your organization and its activities are aligned with your core values?

We have a well established set of core values that labor and management developed together. I make it a point to set a high standard for my own conduct and performance. I model those core values by leading by example from the front and I hold my junior officers accountable for similar conduct, as well as mentorship. I encourage the same with their company officers.

What are you doing to ensure you continue to grow and develop as a leader?

I continue to be a lifelong learner. I am always looking for education and training opportunities. Additionally, I look for opportunities to mentor other chief officers, both inside and outside of my agency. Mentoring had a huge impact in my career development but most of my mentors are retiring out of the career field. I find that being active as a mentor with more junior folks keeps me motivated, energetic, and focused on the latest issues.

16

ANDREW LIPARI, CAPTAIN, ARFF TRAINING OFFICER, KCFD

How did you get to your position? What path did your career follow?

I started my service career in the US Air Force in Fire Protection (Jan 87 - Sept 91). I joined the Kansas City Fire Department in March of 93 until the present. I have been involved in, and around, ARFF for over twenty-five years. I have been the ARFF Training Officer for the last four years.

What is your advice for the newly promoted chief officer?

Remember where you came from. Keep learning. Seek the council of those above you and experienced members below.

What was the most impactful call/emergency you have been on? Why?

Non-breathing infant when I was in the Air Force, and only nineteen years old. It was my first real emergency. The infant did not make it. However, watching my superior stay calm, offer instructions, and control the scene helped me see how important it is to know what to do when the scene gets chaotic.

What actions, behaviors, or thought patterns lead to leadership failure?
Thinking that your way is the only way. We seldom have to reinvent the wheel. Getting to the position you desired and thinking you've made it and forget that you still get to learn.

What is the top trait or characteristic that you believe every chief officer must possess?
Listening. Do what is right and fair.

What is a habit or routine that you have and how does it help you persevere?
Set aside some time each day to learn something new. Change is constant.

What have been the key factors to your success and why?
Giving a sh!t! When you care about yourself, your peers, and your job it is easy to succeed.

What are the most important decisions you make as a leader of your organization?
Decisions that center around the safety of myself and others.

How do you ensure your organization and its activities are aligned with your core values?
This is a tough one. The two don't always line up. Be vocal about

what I believe is in conflict with what is right. If a rule needs revisiting then don't just accept that "it's in the rules" excuse.

What are you doing to ensure you continue to grow and develop as a leader?

Learn. Emulate the things I see others do that are right and do what I can to make a difference for the better.

CHRISTOPHER MENGE, CAPTAIN, TRAINING OFFICER, ALBANY AIRPORT FIRE DEPARTMENT

How did you get to your position? What path did your career follow?

I started in the military as a firefighter progressing ten years to the rank of Assistant Chief and then took the position as a firefighter with the airport. After five years I took the promotional exam and scored well enough to be promoted. I have continued to do the same to the rank of Captain that I now maintain.

What is your advice for the newly promoted chief officer?

Trust the officers below you to accomplish your tasks. Realize that you are a team and need to work together in all aspects. Keep in mind that your success is based on those who are following you, be a leader not just in rank but in life.

What was the most impactful call/emergency you have been on? Why?

LODD in the military that should not have occurred if the firefighter had some ability to adapt to the situation that was occurring. I was involved in a large scale training event (live burn) which caused a house to explode and leave eleven firefighters injured. It led me down this path and keeps me passionate about training.

What actions, behaviors, or thought patterns lead to leadership failure?

Failure to take action and hold other people responsible for their actions or lack thereof. Failure to provide discipline and direction will destroy your ability to lead.

What is the top trait or characteristic that you believe every chief officer must possess?

The ability to take ownership of their actions and those of its members you lead.

What is a habit or routine that you have and how does it help you persevere?

I like to continue to focus on education and share that with my peers. It keeps me aware of the changing fire service and allows me to remember how fragile our knowledge base can be.

What have been the key factors to your success and why?

Preparation, focus, and the willingness to adapt to situations and the changing landscape.

What are the most important decisions you make as a leader of your organization?

How you treat your people.

. . .

How do you ensure your organization and its activities are aligned with your core values?

GREAT question. I think we all struggle everyday with defining and maintaining that path. The choices we make, intentionally or unintentionally, have consequences in every aspect of our day. Perception is reality.

What are you doing to ensure you continue to grow and develop as a leader?

Constant education and learning about my people, and their lives, so I understand their decision making process.

* * *

Calderwood Training Solutions
https://www.facebook.com/calderwoodtrainingsoltuions/

ALVIN LEE, CHIEF, AIRPORT EMERGENCY SERVICE, CHANGI AIRPORT GROUP

SINGAPORE AIRLINES FLIGHT 368 RESPONSE

How did you get to your position? What path did your career follow?

I have been in the law enforcement and emergency services field for over twenty years - first as a Senior Police Officer in the Singapore Police Force (SPF) and now, as Chief of Changi Airport Emergency Service (AES). Prior to this present appointment, I had also served as Head Planning, Commander Military, and Commander Civil. During my career in the SPF, I have been in strategic, operational and tactical command roles. In addition, I have also served at the precinct, divisional HQ, Force HQ and at Ministry HQ, as well as in both command and staff positions.

What is your advice for the newly promoted chief officer?

Never underestimate the value of collective leadership.

What was the most impactful call/emergency you have been on? Why?

SQ368 engine fire on landing. While it was a textbook scenario, the follow up investigation process was nerve wracking.

What actions, behaviors, or thought patterns lead to leadership failure?

Complacency and hubris.

What is the top trait or characteristic that you believe every chief officer must possess?

Diplomacy and determination.

What is a habit or routine that you have and how does it help you persevere?

Reading - learn vicariously from the experience and knowledge of others.

What have been the key factors to your success and why?

God's grace (being put in the right place, at the right time), past experience in the police force (good grasp of emergency/crisis management and people skills).

What are the most important decisions you make as a leader of your organization?

Strategic decisions to meet problems of today and the challenges of the future (equipment, manpower, investments, conops factoring long term airport growth, etc).

How do you ensure your organization and its activities are aligned with your core values?

Having a clear mission, vision, and values that guide everyone in the service.

What are you doing to ensure you continue to grow and develop as a leader?
Read regularly, be in the company of people that constantly challenge your mindset and help you grow. Get out of your comfort zone, do different things, and gain new experiences.

* * *

On 27 June 2016, a Boeing 777-300ER aircraft departed Singapore for Milan. About two hours into the flight, the right engine indication showed a low oil quantity and subsequently, the flight crew felt a vibration in the control column and cockpit floor. The flight crew decided to return to Singapore.

Shortly after landing in Changi Airport, a fire was observed to have occurred in the vicinity of the aircraft's right engine. After the aircraft came to a stop on the runway, a fire developed under the right wing. The airport rescue and firefighting service, which was already on standby, responded promptly and the fire was extinguished. All persons on board the aircraft disembarked via a mobile stairs.

Damage to the aircraft included heat damage to the core of the engine, portions of the engine cowlings, the wing area directly behind and outboard of the right hand engine. There was no injury to any person in this occurrence.

The occurrence was classified as an accident.[1]

— TRANSPORTATION SAFETY INVESTIGATION BUREAU, MINISTRY OF TRANSPORT, SINGAPORE FINAL REPORT

What lessons did your department learn?
The investigation into an aircraft incident can be rather stressful for the Operations Commander (every shift is led by an Operations

Commander and is assisted by a Duty Officer and the rest of the crew) and the team. It is extremely important to be able to provide all the necessary details of the incident e.g. the detailed incident log, key timings, etc to the investigators.

In addition, the Operations Commander should be well trained and be able to explain clearly the decisions he made during the incident. In this case, one key question was who had made the decision to disembark the passengers using the air-stairs instead of using the emergency evacuation chutes.

It is also very important to be conscious of the actual verbiage when advising the pilot during the incident.

What changes were made?
Training of Operations Commanders was stepped up to emphasize the importance of clear radio communications during aircraft incidents and the need to maintain composure despite the crisis.

This and other such incidents were also shared in the Operations Commander training including play backs of the actual conversations between Operations Commanders and the pilots in such incidents.

Had your training and experience to this point adequately prepared you for this incident? How, why, or why not?
Fortunately, this was an engine fire incident that is part and parcel of our training. Hence, from an operations perspective, everyone on duty were familiar with the scenario and their roles.

What was perhaps 'educational' for all involved was the scrutiny of the investigation which not all crew had experienced.

What impact has this had on your career and leadership?
The impact from the perspective of career advancement was negligible. However, given the high safety in air travel and the limited

opportunities to deal with aircraft incidents, this was an invaluable experience for my team and I.

This was especially so, as stated above, with regard to working with the investigators to understand fully what had happened on the ground and how we can learn the necessary lessons to further improve air travel safety.

What advice would you give to prepare someone for responding to an incident of this magnitude?

I cannot over emphasize the importance of regular training of the entire team in playing out various aircraft incident scenarios. Such training would involve both physical training with our aircraft simulators, as well as, table top exercises where the responding crew and the commanders talk through the various scenarios.

It is also important to have a separate unit/s (apart from the responding operations crew) to test and evaluate the operations crew on a regular basis. We have an Inspectorate unit that carries out company and individual level proficiency tests of all operations crew annually. In addition to this, we also have an Operations Planning team at HQ who plans and conducts regular exercises (both tabletop and ground deployment exercises) to exercise the operations crew on familiarity with the different emergency plans (e.g. aircraft crash, aircraft crash at sea, hazmat incidents, etc).

It is through such regular exercises and tests that the crew will be able to respond, almost instinctively, to any aircraft incidents we could potentially encounter.

More specifically, my advice to the team that is actually responding to an incident would be to remain calm. Trust that all the training you have been undertaking will kick in and you will be able to respond accordingly – almost second nature. So "train like you fight. Fight like you train!"

CHRIS THAIN, BUSINESS DEVELOPMENT MANAGER, FIRE & RESCUE SERVICES, G3 SYSTEMS LTD

How did you get to your position? What path did your career follow?

Ten years in the Fire & Rescue sector, eight years with a UK Local Authority Fire & Rescue Service at Senior Leadership Team level and two years with a local provider of industrial Fire & Rescue Services. Prior to this I marketed fire, rescue, and safety equipment and services to the maritime industry for eighteen years.

What is your advice for the newly promoted chief officer?

Listen to your staff, be clear in your direction, communicate effectively through all levels of your organisation, don't forget to go out to jobs - sitting in your office is not going to inform you about how well your service operates. Be visible.

What was the most impactful call/emergency you have been on? Why?

A woman had fallen under a high speed train. In difficult circumstances the crews responded as a complete team to access, stabilise and retrieve the individual, who survived. The care, trust and teamwork I saw that day inspired me.

What actions, behaviors, or thought patterns lead to leadership failure?

Group think, failure to act on lessons learned from mistakes, weak and indecisive senior leadership, more focus on political expediency and career management than organisational excellence.

What is the top trait or characteristic that you believe every chief officer must possess?

Honesty and integrity. The ability to listen and understand. A good sense of humour.

What is a habit or routine that you have and how does it help you persevere?

Keeping on top of my filing - so that I can always find what I need, when I need it. Leave work at work whenever possible. I try to focus on my family and home, as they deserve my undivided attention when off work. Try to be a good friend at all times. Treat people as you would like to be treated yourself.

What have been the key factors to your success and why?

Taking ownership of issues when nobody else seemed to want to, because it was the right thing to do - "proceeding until apprehended". Calling out corruption despite it being personally damaging to my career. I can look in the mirror each day knowing I did the right thing. Working hard and diligently at all times. Respecting others in the workplace and always being courteous and good mannered.

. . .

What are the most important decisions you make as a leader of your organization?

Risk assessing every situation and adapting our approach, as required, in order to ensure a safe and successful outcome.

How do you ensure your organization and its activities are aligned with your core values?

Lead by example. Do not be afraid to raise concerns when you see them, even if doing so could lead to personal career damage. Buy into, and live, your organisational values, even when others around you fall short.

What are you doing to ensure you continue to grow and develop as a leader?

Continuing professional development. Membership of organisations such as JOIFF, AFOA, NFPA and IFE enable me to understand developments across the Fire Sector and to network with like minded individuals around the world. I write feature articles for the fire sector media and support initiatives to further enhance fire safety and fire services around the world.

20

THOMAS LITTLEPAGE, CAPTAIN, SHIFT LEADER, EVANSVILLE REGIONAL AIRPORT SAFETY DEPARTMENT

How did you get to your position? What path did your career follow?

Eight years active duty in the Air Force. I came to my current airport as a twenty-five year-old officer who started as a regular airport safety officer.

What is your advice for the newly promoted chief officer?

Make a list of all the things you criticized the former chief's for, and don't do any of them.

What was the most impactful call/emergency you have been on? Why?

It was the first crash I responded to in the Air Force, as a young inexperienced firefighter. I saw the entire process of working a scene from start to finish.

. . .

What actions, behaviors, or thought patterns lead to leadership failure?

This is an easy one because I see it all the time. Being selfish and trying to jam your own vision or agenda down the throat of the department, because you think you are smarter and know best.

What is the top trait or characteristic that you believe every chief officer must possess?

Humility. It's not about you, it's about everyone else. This one trait will lead to being a better communicator, being better organized, thinking about the firefighters under your command. It leads to team building.

What is a habit or routine that you have and how does it help you persevere?

Consistency. Behaving in a consistent manner at all times. I never let issues, or my personal feelings dictate how I act and treat others.

What have been the key factors to your success and why?

All the factors mentioned above. Humility, consistency, and treating others with respect at all times.

What are the most important decisions you make as a leader of your organization?

The small decisions about routine issues show your true character. If you are unfair with small leadership tasks, you will struggle with the bigger ones.

How do you ensure your organization and its activities are aligned with your core values?

Communicate your core values, and what is expected to your department. Be consistent with those values and apply them fairly.

What are you doing to ensure you continue to grow and develop as a leader?

Keep moving forward. So much new technology is involved in aviation, and it seems to be obsolete by the time I master it. Then the next "new" thing comes along every other month.

21

PETER MOORE, MANAGER, AIRPORT FIRE SERVICE, CHRISTCHURCH INTERNATIONAL AIRPORT LTD

How did you get to your position? What path did your career follow?

I started as a volunteer in my community brigade then joined ARFF as a career (continuing as a volunteer for twenty years). Usual progression through the ranks to Fire Officer and acting Senior fire Officer, then overseas to Rarotonga as CFO. Returned to Christchurch as Fire Officer, appointed Senior Fire Officer, Training then promoted to Chief in 2005 on the retirement of the incumbent. I have been a firefighter for a total of forty-six years now, having served forty-two with ARFF and Chief for the last sixteen years.

What is your advice for the newly promoted chief officer?

Have faith in your own abilities, that's what got you here! Gut instinct is real, it comes from recognition-primed decision making. Engage with your subordinate officers, you'll only be as good as they allow you to be.

Do something early in your time to benefit staff, e.g. renovate the mess room, renew gym equipment, replace the lounge chairs, upsize the TV. It doesn't have to be big, but something just for them. You will be surprised at the brownie points you'll get.

What was the most impactful call/emergency you have been on? Why?

First fatal fire as OIC (in my volunteer brigade). House fire, well involved, child reported missing. Located and extracted in quick time but deceased. Feedback from the back-up career crew confirmed we had done a good job and couldn't have done any more. This was a huge relief, obviously, and a confidence boost in my ability to continue in that role.

What actions, behaviors, or thought patterns lead to leadership failure?

Not being fully committed to a course of action. Doubting what you are doing is right even when directed to by superiors and not having challenged that strongly enough (in the appropriate manner/setting). This will be perceived as disingenuous.

What is the top trait or characteristic that you believe every chief officer must possess?

Being true to himself. Acting with honesty and integrity. Displaying confidence.

What is a habit or routine that you have and how does it help you persevere?

Having a set start time. For me flexible hours/glide time lead to procrastination. I'm always on station before morning shift change (0800). This not only allows me to connect with the outgoing team and be involved with the incoming morning parade but shows I'm not "above the law".

. . .

What have been the key factors to your success and why?

Knowledge - I got qualified, and then some, and continued learning.

Connections - I took the opportunities that arose and made a point of establishing an ongoing network with sister organizations and key suppliers.

What are the most important decisions you make as a leader of your organization?

We are an emergency service in a corporate organization (civil airport) so maintaining our purpose of emergency response against pressure to do more non-emergency routine tasks that may compromise response can be challenging. Everyone seems to know our job better than us so drawing a line in the sand, even if that means being bloody minded sometimes, is important

How do you ensure your organization and its activities are aligned with your core values?

Walking the talk. Leading by example so that when poor behaviour or performance needs addressing you can do so from a place of integrity.

What are you doing to ensure you continue to grow and develop as a leader?

I'm nearing the end of my career now so I am no longer pursuing formal education, but I am taking the opportunities to attend conferences, seminars, and refresher training wherever possible

22

LTCOL. FRANCOIS VILLARD, OWNER/GENERAL MANAGER, AIR SAFETY AND SECURITY SERVICES

How DID you get to your position? What path did your career follow?

I was previously a researcher in paleontology and hold a master's degree in paleontology from the Sorbonne University in Paris. I became the second in command at the Geneva airport in 1988. Then I was upgraded as chief of the Airport Safety and Security Department and chief of the Emergency Command Post in case of aircraft incidents. Simultaneously, I was a militia army officer and I commanded a battalion of tanks. I was professionally trained by the International Fire Training School. I was an expert of the Swiss civil aviation and a trainer in the three national languages for airport firefighters for six years. In total I have completed nearly fifty weeks of aircraft fire training on various training grounds but mainly in Teesside (UK). I had very close relations with Swiss airlines and provided several training sessions for these companies.

I had the opportunity to make many flights as an observer in the cockpit of airliners. Thus, having organized the emergency exercises of the Swiss airports, I always integrated these airlines as extras. In charge of the command of the rescue and coordination operations in case of air crashes, I was the almost immediate observer of fourteen air crashes (Zurich airport, Amsterdam, Strasbourg, Paris, Turin, Toulouse, the Germanwings case in the Alps). Using the lessons learned from these air crashes, I developed a concept of advanced medical post loaded on a container, developed a cradle for aircraft de-sludging after having participated in several aircraft recovery operations. Following the Strasbourg accident, I organized SAR search exercises and did survival courses on land and water with the airlines.

In the field of airport security, I was in charge of the training of passenger controllers, made several interventions for bomb alarms, participated in the intervention on a hijacking, was the architect of a new design of the screening points, the control of hold luggage, cargo and catering. I was personally trained and integrated in a team of improvised bomb defusers.

I was admitted to the NFPA aviation working group in 2000. I started my own consulting training company and continued to train many firefighters in Europe, the Middle East, the Pacific and Africa. By accompanying department chiefs, I ensured their support during airport incidents.

What is your advice for the newly promoted chief officer?
The new leader takes on a load that can absorb him completely. The family must be prepared for this. The area of fire and intervention represents 5-10% of the chief's activity for which he must be completely prepared both morally and technically. The rest is mainly represented by personnel management, command, and an emphatic approach to personnel problems.

It is also necessary to have a well-made head who knows how to focus on the essential things and leave out the superfluous. In his command activity he must leave room for subordinates to exercise their power to lead their teams without interference from the hierarchy. In cases of disputes, he must know how to analyze the facts impartially and sanction them when necessary. The leader certainly exercises the power to sanction, but he must never forget the power to congratulate and encourage.

As far as his function is concerned, he must lead with a forward-looking vision of things, establish sensible and well-framed budgets. With his direct hierarchy, do not forget that his service, because of its cohesion and the wearing of the uniform, presents a message that some directors fear or lead from a distance.

What was the most impactful call/emergency you have been on? Why?

I have witnessed many accidents and it is of course always a highlight when you see, relieve and accompany, if necessary, the victims. For me, one of the highlights was a perfectly successful intervention on a landing with the landing gear retracted on a foam carpet (that was possible at Geneva Airport). We did this under many watchful eyes, as journalists were expecting Brian Jones and Bertrand Piccard back from a hot air balloon around the world, they watched this intervention with their noses glued against the windows of the airport terminal.

It was indeed the crowning achievement of an extraordinary cooperation with the tower controllers (with whom I have always maintained close relations and collaboration), with the airline company, and the pilot.

What actions, behaviors, or thought patterns lead to leadership failure?

Probably an incompetence in command, that is to say a lack of decisiveness, an inability to make situational assessments. I have sometimes observed leaders who intervene in the subordinate hierarchy, demotivated by preferences between sappers and subordinate leaders. I think that a leader must absolutely have legitimacy and technical and human professional recognition, otherwise his command fails. The leader must be an example and if he acts by compromises or weaknesses he cuts himself off from his subordinates.

. . .

What is the top trait or characteristic that you believe every chief officer must possess?

Certainly a confirmed professional competence, a clear and firm attitude. Acting with praise first and before reprimand. Keeping in a good mood, respectively keeping one's private sphere to recharge one's batteries and find energy.

A chief must therefore be well presented, express himself with a precise vocabulary (which can sometimes be quite crude, why not?). Never admit injustice and respect his collaborators and subordinates, leaving them free to carry out their activity and their mission as a leader.

What is a habit or routine that you have and how does it help you persevere?

Be well organized, use the skills of your subordinates, and respect the time. Inform as much as possible, both in daily work and in his activity (use his means of communication to tell what is happening, place steps, etc.).

What have been the key factors to your success and why?

Above all, experience, motivation to the point of excess, and fidelity to one's work and mission.

What are the most important decisions you make as a leader of your organization?

Difficult question. Probably to keep an unwavering line of conduct in its objectives. In my professional career there have been many decision-making crossroads that have committed the future. By having long term objectives we know how to make these crucial choices and I encourage future leaders to sometimes take three steps backwards, stop looking at the handlebars, and just think about the future.

. . .

How do you ensure your organization and its activities are aligned with your core values?

By my personal satisfaction, by my mood, and by the relationships I have with my partners. These people recognize me as competent and sometimes admiring of my achievements.

What are you doing to ensure you continue to grow and develop as a leader?

I'm always reading up on myself, I like to take part in all kinds of initiatives and professional activities.

23

ELIZABETH HENDEL, DEPUTY FIRE CHIEF (RET.), PHOENIX FIRE DEPARTMENT

How did you get to your position? What path did your career follow?

On January 3, 1983 I started my career with the Phoenix Fire Department, as the eighth female hired into a predominantly male workforce. As the first female to go on "B" shift, I took every promotional opportunity afforded me. I was a firefighter on the trucks for approximately seven years before being promoted to Engineer (driver), a position I held for eight years. I then promoted to Captain and on to Battalion Chief. Deputy Chief came a year later when I was again promoted and took over Sky Harbor International Airport.

I had the fortunate experience to serve with great organizations such as: Aircraft Rescue Firefighting Working Group (ARFFWG), National Fire Protection Association (NFPA), International Fire Service Training Association (IFSTA), National Academy of Sciences (NAS), along with several others. I was given the opportunity to serve on two committees writing fire service manuals for Chief Officer and ARFF. In addition, I served as the Chairperson to write and publish an after-action report for the National Transportation Safety Board on the avoidance of striking passengers from the Asiana 214 incident in San Francisco on July 6, 2013.

I was able to travel internationally and speak in several parts of the world on behalf of the ARFFWG as the Chairperson. Being the first female to serve as chairperson for this international organization was an honor and extremely rewarding. I have been retired officially from the American Fire Service and instruct at the local community college as well as in the ARFF industry. This in a small way helps to give back to a full career which has afforded me so many opportunities.

What is your advice for the newly promoted chief officer?

Listen to your people. Believe in them and support them. Give them the tools to do the job, then get out of grown peoples way to allow them to do their job. Don't compromise your moral values or your principles. Always strive to do the right thing and you certainly can't go wrong. Never compromise legal matters, uphold the laws so it doesn't bite you or your department. Ask for help from your mentors, your personnel department, and your union if you need to. We are one team.

What was the most impactful call/emergency you have been on? Why?

I recall the first fatality I responded to. It was a bad accident with four patients lying in the street. It occurred close to our station shortly after 1 a.m., soon after the bars had closed. Two vehicles at a high rate of speed had collided throwing the passengers throughout the streets from their vehicles. As we worked hard to get the patients treated and transported and leave those who, unfortunately, did not survive we packed up our gear and headed back to the station. We cleaned up and restocked everything for the next call and hit our bunk beds waiting for the next alarm. As I lay there attempting to fall back asleep, Captain Tucker came into the dorm room and called out my name. We left the dorm room so the other firefighters could get some rest. Captain Tucker had me go out and sit in the fire truck across from him as we discussed the call we just went on. He was genuinely concerned for me and wanted to make sure what I had witnessed was not going to affect me emotionally. As I recall, we talked for an hour or so, he had been on the job for a while and had seen much worse. Captain Tucker was ahead of his time. He had given me great advice and welcomed me to talk about calls with him anytime. Discussions like these would not occur on our job for another twenty years until Post Traumatic Stress Disorder would be recognized in the fire service. I am grateful he was my first Captain; we would later work together when I was pregnant with my first daughter and he was a division chief in Public Affairs. Captain Tucker went on to be a Fire Chief in New Mexico.

What actions, behaviors, or thought patterns lead to leadership failure?

Thinking the buck stops with you. You need to lead by example and trust in your people. Always do the right thing. It's when you don't do the right thing, things get out of sorts. Never do knee jerk reactions. Ask for help when you need it.

What is the top trait or characteristic that you believe every chief officer must possess?

Don't be afraid to make a decision. You have to make hard decisions as a chief, life and death ones as a matter of fact. Make the decision based on knowledge, SOP/SOG's, ethics and laws, and past history. You need to be approachable and be there to serve the members. Reward good work too and let everyone know you appreciate them.

What is a habit or routine that you have and how does it help you persevere?

Start your day out with thought and meditation. Be prepared for your day and make the most of it. Always take advantage of seminars, webinars, and educational opportunities to keep you growing and learning. Times change, change with them

What have been the key factors to your success and why?
Staying active in the department and outside the department in order to stay relevant in the fire service.

What are the most important decisions you make as a leader of your organization?
Safety, both on and off the fire ground. keeping everyone motivated to stay healthy and strong in order to be prepared for the next call. You have to make sure everyone goes home safe and sound at the end of the day to, hopefully, retire healthy and happy - both physically and mentally.

Additionally, standing up for what is right.

How do you ensure your organization and its activities are aligned with your core values?
You are the leader, lead by example and instill these values in your department and section. Help in the writings of your policies to ensure these are good values to uphold.

What are you doing to ensure you continue to grow and develop as a leader?
Staying active on key committees and relative in the service until one day I decide to fully retire,

Treat people with respect until they prove otherwise. Take opportunities and surround yourself with positive people who help you to be a better person. Early on in my career a chief whom I looked up to once told me, "Tell me who your friends are, and I'll tell you what you're like". Being very young at the time it was a hard concept to follow, but I took his advice and changed a few people in my life who were bringing me down. This is still one of my favorite quotes to this day.

Stand up for what is right. Give everyone a chance and do your job to the best of your ability. You cannot go wrong when you do the right thing. People will respect you for standing up for what is right, you can be that one person who changed things for the better. I learned this from the captain who did not want me to drive for him; when you are wrong, admit it, apologize, and move on.

Take the opportunity when it comes your way. You may not want to be in a certain place at the time, try to make the best of it, it just might work out. Going to the airport turned out to be a positive move for me. It led me into a career path that is my passion. As a chief officer, I would counsel members who resisted change by telling them my story of fighting the change to go to the airport. People in general do not like change, but firefighters really don't like it! They will either tell you, "Tradition is too strong, and that tradition is important", or "It's always been done this way". Change can be good, as it is necessary for growth.

Always do your job to the best of your ability. Ever since I was a little girl, my father would say, "Always do your best". This was one of three in his almost daily advice, the others being, "Everything in moderation", and "don't complain, nobody wants to hear it". Advice still worthy of living by today.

Mentor others and share your knowledge to help them become better, one day they may be your replacement or your boss. As in the story of the captain going home, don't be forced into picking a fight that is not yours. Have friends who will warn you when you are being placed in a hostile environment, this way you are able to make informed choices. Do not take things personal, unless they are, know the difference.

ANTHONY DYNDERSKI, FIRE CHIEF, SIKORSKY AIRCRAFT

How did you get to your position? What path did your career follow?

Out of high school I worked for a fire equipment and truck dealer, O.B. Maxwell. The business sold Maxims and Pierce. At that time, Pierce sold Commercial Apparatus. I met a number of fire service leaders who I got to know. I started in a very active volunteer company in the early 70's. The department had a documented firefighter 1 and firefighter 2 program. I became an EMT in 1975 and am still certified today. In the early 80's I started my degree program at the University of New Haven. It took some time, but I eventually completed, and have, a Masters in Emergency Management and Graduate Certificate in Public Safety Management. I worked for the Connecticut Fire Academy as an adjunct instructor specializing in industrial fire protection and AFFF. I taught Special Hazards at the University of New Haven. My best job was a shift officer at Sikorsky!

What is your advice for the newly promoted chief officer?

Be and act like an officer. If one does not like standing up and doing what's right, don't become an officer. Act consistently.

. . .

What was the most impactful call/emergency you have been on? Why?
Dispatching three firefighters to the Wall Street Heliport on the morning of 9/11 prior to the towers falling. Why - I sent three people into harms way. It wasn't just a fire, it was a terrorist attack!

What actions, behaviors, or thought patterns lead to leadership failure?
Not doing one's job, inconsistent, not continuing with education. Falling victim to harassment.

What is the top trait or characteristic that you believe every chief officer must possess?
Tenacity- never give up, keep pushing forward, making improvements. Take risks. Investigate and adopt new practices, procedures, and equipment.

What is a habit or routine that you have and how does it help you persevere?
I never give up and I don't let anyone get me down. Walking through the apparatus bay and seeing what I accomplished keeps me going.

What have been the key factors to your success and why?
Never giving up.

What are the most important decisions you make as a leader of your organization?

Staffing, safety, budgeting. Taking on the administrative role and risk management for emergencies.

How do you ensure your organization and its activities are aligned with your core values?

I take an active role. I talk to people and mentor. Lead by example. My goal is discipline and I don't mean punishment. I am trying to get everyone to do the same things all the time.

What are you doing to ensure you continue to grow and develop as a leader?

Read! Take classes, use the internet.

DANNY PIERCE, AIRPORT SAFETY OFFICER (RET.)

How did you get to your position? What path did your career follow?

Thirty-six years of ARFF - including thirty years with Los Angeles World Airports and four years with the United States Air Force.

What is your advice for the newly promoted chief officer?

Always agree publicly with superior officers.

What actions, behaviors, or thought patterns lead to leadership failure?

Lack of training, not pursuing education.

What is the top trait or characteristic that you believe every chief officer must possess?

Listening to lower ranking personnel, demonstrating leadership skills.

. . .

What is a habit or routine that you have and how does it help you persevere?

Constantly use the internet to stay updated on new operational techniques.

What have been the key factors to your success and why?

Education used with experience. Also, learn from the mistakes of others.

What are the most important decisions you make as a leader of your organization?

To follow SOP's with training and keep crew safety in mind at all times.

What are you doing to ensure you continue to grow and develop as a leader?

Education and NFPA participation.

* * *

ARFF Solutions
www.arffsolutions.com

26

GRAEME DAY, FIRE SERVICE ASSURANCE MANAGER, CAPITA FIRE AND RESCUE

How did you get to your position? What path did your career follow?

Joined the structural fire and rescue service and served for thirty years, retiring as an Operational Area Commander. On retiring I was appointed as the Fire Service Oversight & Regulation Manager covering seven UK airports based at London Heathrow Airport. I joined Capita Fire and Rescue in November 2019 as the Fire Service Assurance Manager covering the UK Military, ARFF based.

What is your advice for the newly promoted chief officer?

Be accessible to your staff, represent the service professionally, be prepared to make tough decisions and be humble.

What was the most impactful call/emergency you have been on? Why?

Fatal multiple vehicle accident involving minors. It made me think about how fragile life is.

. . .

What actions, behaviors, or thought patterns lead to leadership failure?

Not communicating effectively and not trusting colleagues.

What is the top trait or characteristic that you believe every chief officer must possess?

Clear leadership.

What is a habit or routine that you have and how does it help you persevere?

Having a clear plan that keeps me focused.

What have been the key factors to your success and why?

Commitment, willingness to learn, ability to listen, trusting colleagues.

What are the most important decisions you make as a leader of your organization?

The ones that affect the core objectives of our organisation.

How do you ensure your organization and its activities are aligned with your core values?

By being aware of my values and behaving in a way that ensures they are reflected in the day to day workings of my organisation.

What are you doing to ensure you continue to grow and develop as a leader?

Continually listening to colleagues and senior managers.

PETER MCMAHON, MANAGING DIRECTOR, AVIATION RESCUE SERVICES

How did you get to your position? What path did your career follow?

Recruit Firefighter to Assistant Chief Fire Officer (twenty years industry experience). Lots of hard work, and always sought opportunities to learn and contribute to the service.

What is your advice for the newly promoted chief officer?

Empower and equip your direct reports to be better than you! Lead by example and never skimp on SAFETY! Always place the health and safety of your team as the number one priority. Have robust change management and quality assurance and validation processes. Resource your training department.

What was the most impactful call/emergency you have been on? Why?

Full terminal emergency involving evacuation of more than 5,000 people. A complex incident involving HAZMAT, multi-agency response, and the need to prioritise life safety over business continuity.

What actions, behaviors, or thought patterns lead to leadership failure?

Complex question, but leadership without a mission or strategy is like a ship without a rudder. Failure to delegate (and/or allow your team to lead and resolve issues), ATTITUDE!!

What is the top trait or characteristic that you believe every chief officer must possess?

Integrity
Ability to delegate
Good communicator
Self-awareness (awareness of strengths and weaknesses)
Gratitude
Learning agility (always seeking to learn)
Influence
Empathy
Courage
Respect

What is a habit or routine that you have and how does it help you persevere?

Chatting informally with firefighters every day. A chief should not seem aloof or above his team.

What have been the key factors to your success and why?

Perseverance, continual self development, and effective networking.

. . .

What are the most important decisions you make as a leader of your organization?

Many, but implementing policies and procedures and ensuring staff are safe.

How do you ensure your organization and its activities are aligned with your core values?

Regular meetings with team to ensure we are on track, key performance measurables, internal and external feedback.

What are you doing to ensure you continue to grow and develop as a leader?

Continual self development and evaluation, mentor program (I have an external peer who provides support, advice, and mentoring).

28

ROB MATHIS, ASSISTANT FIRE CHIEF, PORTLAND AIRPORT FIRE AND RESCUE

How did you get to your position? What path did your career follow?

I started my career as a volunteer Firefighter/EMT while attending the fire academy and completing my Associates Degree in Fire Science. Upon graduation, I joined the USAF as an Aircraft Rescue Firefighter. After four years and an honorable discharge, I was hired as a Firefighter/EMT with Boeing Fire Department. I spent the next twenty-six plus years there in the roles of Firefighter, Lieutenant, Captain, BC, AC and finally, Fire Marshal. In 2018 I left Boeing Fire and joined Portland Airport Fire and Rescue as their new Assistant Chief.

What is your advice for the newly promoted chief officer?

Do not forget where you came from. Remember what it was like to be that firefighter on the floor. Make time to just sit and visit with your crews. Maybe have dinner with them once a week. Be a servant leader and never forget that attitude reflects leadership. Be a positive influence.

. . .

What was the most impactful call/emergency you have been on? Why?

Not the most exciting call I have been on, but it was one of the first. In my first two weeks as a volunteer firefighter I had the opportunity to do CPR. In both cases the patient did not survive. However, both of those calls have remained with me in great detail. I believe these two responses solidified my career choice in emergency services. Just seeing and hearing the desperation in the eyes and the voices of the loved ones at the scene, and knowing they were counting on us to make a difference really hit me. I decided that I wanted to make a difference from that point forward. I have never looked back or second guessed my career in more than thirty-four years.

What actions, behaviors, or thought patterns lead to leadership failure?

Forgetting where you came from and how you got there is a recipe of failure. Your department is only as strong as the weakest leak and it's your responsibility to identify opportunities to improve your team. Never stop learning and always be willing to listen. The fire service is a dynamic environment that, as a leader, you need to evolve and grow with. Champion your team and surround yourself with the best. Do not be intimidated to promote people that are more talented than you. Always put the team and department above self. Be a good communicator and an even better listener. Some of the best ideas come from the floor. Empower your people and they will take care of you. Be a servant leader. You are doomed to fail if you don't listen, don't share, stop learning, and make it just a job.

What is the top trait or characteristic that you believe every chief officer must possess?

There are several traits and characteristics that make a successful chief officer but at the top of my list is "integrity". Integrity is not something you learn, it's something leaders are born with. If you have integrity you are an honest and ethical person, not just at work or on the emergency scene, but in every aspect of your life.

What is a habit or routine that you have and how does it help you persevere?

For me it's working out. My day starts with a workout Monday-Friday. This is the only time that I have completely to myself. It helps me to wake up, contemplate and level set for the day. Make time for yourself to help you slow down and start the day fresh. We expect our firefighters to maintain their physical condition and we as leaders need to lead by example. It's a win/win.

What have been the key factors to your success and why?

No leader can make it on their own and if they feel they have then they're not a servant leader. I owe my career to the relationships I have made over my thirty-four year career. Those relationships for me are defined as confidants, coaches, and mentors.

Confidants are close colleagues. These are the people I built relationships with early in my career. Some have advanced through the ranks with me but many remain firefighters. I'm the closest to these individuals, they remind me of where I came from, and are not intimidated to tell me how they feel and what they see. Right or wrong.

Coaches for me are not in a direct supervisory role. They are someone I can turn to to get their opinion on a situation or topic and they can provide perspective and direction.

Mentors for me are usually of a higher rank and in some cases have been my supervisor. They are someone to respect and aspire to lead like. They see something in you and have a vested interest in your success and growth. Again, a good leader is a good communicator and an even better listener.

. . .

What are the most important decisions you make as a leader of your organization?

Personnel is the foundation of your department and will influence the culture of your department. The worst thing you can do is select personnel that are a reflection of yourself. Be diverse in your hiring. Build a team that reflects the community you serve and values integrity and you will have a solid foundation.

How do you ensure your organization and its activities are aligned with your core values?

Listen, really listen. Talk about the core values and believe in them. If your team knows this is a lifestyle and not just words on a wall they will believe in them and you as a leader. Take time to just visit in a non-structured setting. I stay and have dinner once a week and just visit as a member of the team. These dialogs are invaluable.

What are you doing to ensure you continue to grow and develop as a leader?

At this point in my career it's important that I give back. I do this by being a coach and a mentor to individuals both inside the fire service but also outside. Some of the best learning relationships I have is with my friends in Police, Dispatch and Emergency Management. I get just as much out of the relationship as they do. Additionally, being involved with the development of industry publications helps me continue to learn, develop and grow.

PAUL LOONEY, SALES REPRESENTATIVE, PAST CHAIRMAN, ARFF WORKING GROUP

How did you get to your position? What path did your career follow?

Began October 1,1961 USAF CFR (ARFF) until February 5, 1965. Then to Winchester Factory Fire Dept., December 1965 to February 1967. Retired in July 1997 as Lieutenant. Joined ARFF Working Group. Went to South America to increase membership. Formed alliances with NFPA, ALPA, and various fire academies. Went on to work as an instructor at FETF in Ocala, FL, Crash Rescue Equipment in Dallas, and currently serve as a Sales Representative.

What is your advice for the newly promoted chief officer?

Get educated. Go outside your own department to learn the industry (college courses in public administration, fire science, etc.) The further away you get from your own department the more you will be respected by others. Opportunity will find you and you will become more motivated. Your superiors have nothing to gain if you become smarter than them.

. . .

What was the most impactful call/emergency you have been on? Why?

We had a four-alarm factory fire in 1985 (February, freezing cold) and the first in Lieutenant had a heart attack and died upon arrival. The factory fire was really going good and with the death of the first in Lieutenant there was a big delay getting water on the fire. It was 5 p.m. and the beginning of our night shift, I arrived on the second alarm. There was still no water on the fire when I got there.

I was not sure what was going on at the main fire ground, but I found a hydrant next to a fence next to the factory property. I asked the Assistant Chief if he wanted me to drag him a line and he said "yes, yes". So we hooked up to the hydrant and dragged a line over the fence. There were about twenty factory workers that just got off and saw us struggling with the hose (there were four of us); they did most of the work getting the hose over the fence and we were able to drag it another 200 feet to the building involved. Anyway, we showed up with a charged line and gave it to the Chief who was waiting for at least ten minutes hoping to get some water on the fire.

There are thousands of fires I went to but this one stands out. We never told anyone about the factory workers that helped us get the hose over the fence and allowed the Chief to think we were miracle workers. We got back to the firehouse after eight the next morning. You never know what is going to happen when you go to work at a firehouse. Fifteen hours in the freezing cold

What actions, behaviors, or thought patterns lead to leadership failure?

Some people in the fire department are "favorite sons". They are not too smart and get ahead by political connections. Some are okay, but some move forward undeservedly.

What is the top trait or characteristic that you believe every chief officer must possess?

Asst. Chief Paul Hienz, in New Haven, seemed to have this under control. He worked with us. When he gave us a direction (such as break that window, go up the ladder a few more steps, etc.). If we did what he wanted he would encourage us and say, "Yes, that's what I want, keep doing that", "Okay nice job, come down here and rest for a few minutes", things like that.

Some Chiefs give you direction and then say nothing until you make a mistake. This is why we never worried about anything when Chief Hienz was directing a fire operation. We knew he would not ask us to do anything that would get us hurt. He would pay attention and stay in touch with the people putting the fire out.

What is a habit or routine that you have and how does it help you persevere?

Don't trust the leadership if they can't give you a straight answer. That's what Husband Kimmel did at Pearl Harbor. Go to your contacts outside the inner circle of your department. Politics plays a role in municipal departments. Sometimes the guy that moves up the ladder is not the sharpest knife in the drawer.

What have been the key factors to your success and why?

Going outside your own department, attending conferences, networking, and meeting your peers away from where you live. The further you go the more you discover.

What are you doing to ensure you continue to grow and develop as a leader?

Attend conferences and write articles.

PHILIP DIMARIA, BATTALION CHIEF (RET.), MIAMI-DADE FIRE RESCUE

How did you get to your position? What path did your career follow?

Appointed as a firefighter/paramedic in 1986. Promoted to lieutenant 1992. Promoted to captain 1996. Promoted to battalion chief 2004. Assigned to the airport fire division 2007. Retired 2016.

What is your advice for the newly promoted chief officer?

Be a chief. Do your job. You are no longer "one of the guys". You will have to make difficult, and perhaps, unpopular decisions. Make them. Above all, take care of your people.

What was the most impactful call/emergency you have been on? Why?

Having responded to thousands of calls over the course of my career, it is difficult to identify one specific call that I can say was most impactful. Several do stand out, however. I have responded to three major aircraft crashes over the course of my career. These were Valujet Flight 592 in May 1996, Fine Air Flight 101 in August 1997, and Chalks Flight 101 in December 2005. Valujet and Chalks especially standout because of the number of fatalities in each event. Also, responding in the aftermath of Hurricane Andrew in 1992, as a newly promoted Lieutenant, was definitely very memorable.

What actions, behaviors, or thought patterns lead to leadership failure?

Inability or unwillingness to make decisions. Failure to recognize, or acknowledge, mistakes and personal weaknesses.

What is the top trait or characteristic that you believe every chief officer must possess?

Respect for others. Also, the ability and willingness to make decisions.

What is a habit or routine that you have and how does it help you persevere?

I would get to the office early, usually by 0600. This gave me perhaps two hours to work uninterrupted before the time thieves would hit. Regardless of how the rest of the day went, I had a sense of satisfaction that I had accomplished something.

What have been the key factors to your success and why?

Having the support of my superiors. I was extremely fortunate to work for several great bosses who allowed me to do my job and supported me. They gave me the tools I needed to do my job. My goal was to always make my boss look good. I was equally fortunate to have some of the absolute best people work for, and with, me. These individuals were competent, capable, and conscientious. We didn't always agree on issues but we always respected one another. And like

I always tried to do with my bosses, these individuals always made me look good.

What are the most important decisions you make as a leader of your organization?

Foremost, any decisions that impact the health and safety of our personnel. Additionally, decisions that impact the ability of the organization to carry out our mission. If a decision doesn't impact the health and safety of our members, our ability to do our job, or our ability to serve the public, it's probably not that important.

How do you ensure your organization and its activities are aligned with your core values?

I believe that the mission of the fire service has just naturally aligned with my own values of respect for, and service to, others.

What are you doing to ensure you continue to grow and develop as a leader?

Although I am no longer in a professional leadership role, I will always try to continue to grow and develop personally. I still appreciate the opportunity to attend fire service classes and conferences. I read various trade journals and publications regularly.

JACK KRECKIE, CHIEF OF OPERATIONS, ARFF PROFESSIONAL SERVICES, LLC

How did you get to your position? What path did your career follow?

1975 Appointed to Quincy, MA Auxiliary Fire Department (Civil Defense)

June 1976 - Completed training and certification as NREMT

1978-1980 - Firefighter / EMT at General Dynamics, Quincy, MA, Shipbuilding Division

Jan 1980 - Appointed Firefighter / EMT (Crash Crewman) Massport Authority Fire Rescue, Boston Logan International Airport

1985 Promoted to Captain, assigned to the Marine Unit

1992 Promoted to Assistant Chief / Division Commander, Massport Fire Rescue

2005 Promoted to Deputy Chief / Chief of Operations / Boston Logan International Airport, L.G. Hanscom Airfield and Worcester Regional Airport

November 2007, Retired

November 2007, formed ARFF Professional Services, LLC

2008 - 2010, hired as Consulting Employee, (ARFF SME) Science Applications International Corporation (SAIC) assigned as Author / Executive Producer for the FAA ARFF training series.

2010 - present, Contractor (ARFF SME) to support ARFF research programs for the Airport Technology Research Development Program, at the William J. Hughes Technical Center, Atlantic City New Jersey

2012 - Appointed as Global Chief / Aviation Fire Protection - Hostile Environment Services, Perth Australia

2013 - Appointed as ARFF Chief, Komo Airfield, Southern Highlands, Papua New Guinea

What is your advice for the newly promoted chief officer?

Don't forget where you came from. Take care of your people. The day you think you know it all, is the day you should put in your papers.

What was the most impactful call/emergency you have been on? Why?

Serving as the Fire Department representative at the Family Assistance Center at Boston Logan International Airport from 9-11-2001 until 9-18-2001. This was a gathering point for the relatives and airline families who lost loved ones on United Flight 175 and American Flight 11. My eyes still fill up from the stories I heard. The most difficult part for me was working with the child psychiatrist, Dr. Robert, whose job it was to provide for the notification to young children that their parents were not coming home. He liked having me (as a uniformed fire officer) in the room during some of his "talking sessions". He called me his "secret weapon". He said, "look around the room through the eyes of a child, who would you want to hang out with?" There were police uniforms, military, and lots of men in suits. He said, "having a friend who is a fire chief is cool."

I had a second opportunity to work with Dr. Robert at the Family Assistance Center at Otis Air Force Base, with children that had been evacuated from New Orleans from Hurricane Katrina.

. . .

What actions, behaviors, or thought patterns lead to leadership failure?

Failing to lead, because you are trying to remain everybody's best friend. Failing to listen. Failure to admit when you are wrong. Taking credit for the achievements of others. "Don't expect them to step up next time you need a project done."

What is the top trait or characteristic that you believe every chief officer must possess?

Integrity!

What is a habit or routine that you have and how does it help you persevere?

I am blessed with amazing family and friends. Having so many people in my life that I love and trust is critical to maintaining my balance in life. Having someone reach out to me, so I can "talk them off the ledge" (figuratively) bolsters my ego. Knowing that someone trusts me and respects my judgment, that much, leaves me humble. When that person calls back to say "thank you, it helped", is the greatest feeling in the world. I make it a point to stay in touch with people I care about. I love to send cards, notes, and leave messages to say thank you, or I'm sorry to hear. In return, I have people to share my thoughts, fears and self diagnosed shortcomings with. I wake up every morning with my wife (best friend) at my side and look forward to every day. Sharing a video call with any of my three grandchildren puts a smile on my face no matter what kind of day I'm having.

What have been the key factors to your success and why?

That would depend on your definition of success. Every time I took a promotional exam, I studied from the moment the study materials were posted until I picked up the pencil to take the test. My goal was simply to score higher than anyone else. I was promoted after every exam I took. I guess you could say that method was successful, however that is not how I choose to measure my success. I did not succeed in every aspect of my career or my personal life. I did however try very hard, every step of the way. Today, I feel successful and, why not? I continue to get amazing opportunities to remain engaged in ARFF. I love the work and they even pay me to do it. I'm happily married, love where we live, have two amazing families. "My family" are those I'm related to, through blood or marriage. Then there is my "chosen family" of people from all around the world who have blessed me with a bond of mutual love and respect. Those families are evidence of my success in life!

What are the most important decisions you make as a leader of your organization?

My current "organization" is a small "Mom and Pop" ARFF consulting firm. My wife Pat runs the business, manages contracts, keeps us legal, and keeps me on track. Together we decide what projects to pursue or accept and which to decline. We make sure that the scope of work is well within our wheelhouse for every project we take on. The project does not leave our desk until we are comfortable that it is our very best effort. Because of that, we are proud of every project we have completed.

How do you ensure your organization and its activities are aligned with your core values?

We strive to maintain the highest level of integrity. We have turned down projects where the client explained what their goals were and offered to pay us to develop a report that reflects their goals. There are plenty of consulting firms you can pay to provide you with the outcome they desire, but we will not unless we find it to be the best solution. We are driven by safety for first responders, safety of the flying public, and for the airport community. We will not sacrifice integrity for any amount.

. . .

What are you doing to ensure you continue to grow and develop as a leader?

I simply try to do my best every day. When I get an "atta-boy" for something I wrote or a presentation I delivered, from someone I respect, I feel like maybe I've still got game. Being asked to contribute to this book by its author is flattering.

32

ANTONIO GUTIERREZ, A.M.F., P.E.M, FIRE CHIEF, GERALD R. FORD INTERNATIONAL AIRPORT FIRE DEPARTMENT

How did you get to your position? What path did your career follow?

I started as a part-time firefighter in the City of Norton Shores, MI in 1998. I was employed as a part-time firefighter for 20 years (resigned/retired 2018). I was hired full-time as an Aircraft Rescue Fire Fighter in 2008 by the Gerald R. Ford International Airport Fire Department. I was then promoted to Captain in 2016 and finally Chief in 2020. I have served on the County Hazardous Materials Response Team since 2000 and currently serve as the Hazmat Chief. I'm a Professional Emergency Manager along with being a Private Pilot since 2011. I have been an instructor in the fire service since 2004 and have taught at many fire academies, the State Hazmat Training Center (Hazmat Technician Course), and provided safety and hazmat training to the general industry.

What is your advice for the newly promoted chief officer?

Listen more than you talk. Take time to learn about your people. Praise in public, reprimand in private. Set expectations and goals for your people and provide them the tools and resources needed to complete the tasks. Lead by example. Be organized and ready to multi-task. Have confidence. Don't be afraid to ask questions, communication is key. You lead people and manage things. Your greatest asset is your people - don't forget that! Be fair. Hold your people accountable. Never stop learning. Work hard, but also take time to relax. You must have tact when dealing with people. Be prepared and always have back up plans.

What was the most impactful call/emergency you have been on? Why?

I can't say there is one call or emergency that stands out to me. Together all of the calls have impacted me in many different ways. With the death I have seen and dealt with I have learned to appreciate life and each day I'm on this earth. Seeing families go through loss of a loved one or a house, reminds me to make my family the number one priority in life and to appreciate what we have built together. Working with my fellow firefighters over the years on calls has shown me that the fire service is truly a brotherhood/sisterhood. We take care of each other in and out of work. Being able to train the young probies and mold them into successful firefighters is a true honor.

What actions, behaviors, or thought patterns lead to leadership failure?

Not being organized. No confidence. Integrity violator. Zero passion. No commitment. Bad communicator, indecisive, no tact. Zero enthusiasm. Selfishness. No loyalty. Bad judgement. Uneducated. NO ACCOUNTABILITY. Not empowering others, no setting expectations. Not setting goals. Not providing the tools and resources for your people. Not being innovative. No empathy. Not having humility.

. . .

What is the top trait or characteristic that you believe every chief officer must possess?

Inspiration, as a leader, you must be able to inspire your people to complete tasks, follow you, and most of all, to become future leaders. That is the job of a leader, to build future leaders.

What is a habit or routine that you have and how does it help you persevere?

Staying organized. As a leader, you have to stay organized. If you are not organized, you will lose focus. Once you lose focus, everything will fall apart.

What have been the key factors to your success and why?

Leading by example. Staying current in my trade. Taking classes as often as I can (always learning). Earning a bachelor's degree. Watching others and learning what traits are good ones to have. Working hard, being dedicated. Being involved in the fire service. Becoming an instructor and teaching as much as I can. Networking. Reading and learning as much as I can about leadership while also applying it.

What are the most important decisions you make as a leader of your organization?

Anything that deals with my personnel, the budget, and the department image. All of those items could have an effect on me as a leader. Personnel are the organization's greatest asset. I have to make sure they are equipped and prepared to do the job. I must make sure I have the budget to provide the best equipment for my personnel and make sure we have everything we need to do a great job. I must also make sure I have SOP's, SOG's, and policies in place to make sure we, as an organization, conduct our business in a safe, effective, efficient and professional manner.

I also use the acronym PLUS:

- Policies- Are my actions consistent with my organization's policies, procedures, and guidelines?
- Legal- Is the action acceptable under applicable laws and regulations?
- Universal- Does the action conform to universal principles and values adopted by my organization?
- Self- Does the action satisfy my personal definition of what is right, good, and fair?

How do you ensure your organization and its activities are aligned with your core values?

Incorporate the core values, mission statement, vision statement, and motto into everything we do in and out of the fire station. That means living by our mission statement, instilling the core values during training and interactions with the public, focusing on the vision statement and moving into the future. They are not just words, they are what we live by!

What are you doing to ensure you continue to grow and develop as a leader?

I like to learn as much as I can about leadership, therefore I read many books on leadership. I also continue to take courses involving leadership. I have also found teaching the fire officer curriculum in the fire officer academy has allowed me to grow as a leader. Mentoring new fire officers has also helped me grow as a leader.

GARY BARTHRAM, CHIEF FIRE OFFICER, AIRPORT FIRE & RESCUE SERVICE, LONDON HEATHROW AIRPORT

BRITISH AIRWAYS BOEING 777-200ER RESPONSE

How did you get to your position? What path did your career follow?

Started in ARFF in 1989 as a firefighter, worked at two airports before transferring to LHR in 1994, progressed through ranks of Leading Firefighter, Watch Manager, Station Manager, Assistant Chief, Deputy Chief and then to Chief in 2017.

What is your advice for the newly promoted chief officer?

Listen to your colleagues, never be afraid to ask questions or to take feedback. Be sure to interact with all crews when possible so a team ethos can be built. This lets your crews know they have a leader who they can trust and rely on.

What was the most impactful call/emergency you have been on? Why?

Aircraft Accident at LHR in 2008. British Airways 777-200 crash landed just short of threshold due to loss of power to engines on approach. It was the first real major crash I attended and a true realisation of why crews train week in week out for these kinds of events that happen rarely. Training kicks in, giving confidence in actions.

What actions, behaviors, or thought patterns lead to leadership failure?

Complacency, not knowing your subjects and regulations, not listening, and acting before standing back to think things through. A calm measured approach is required whether dealing with emergencies or relations with stakeholders and crews.

What is the top trait or characteristic that you believe every chief officer must possess?

A calm, considered personality which allows you to absorb information before acting on your thoughts. Showing leadership gives confidence to your teams and seniors.

What is a habit or routine that you have and how does it help you persevere?

I try not to jump to conclusions, but weigh out the options or actions before acting on a subject, also not to get over emotional on any subject. When dealing with industrial relations this is particularly important.

What have been the key factors to your success and why?

My technical knowledge in ARFF, my ability to engage with others to gain trust, whether that's my seniors or my crews.

. . .

What are the most important decisions you make as a leader of your organization?

Strategic ones which affect the direction of the department and the crews within it. Operationally, making sure the advice I'm given from my deputy and assistant are thought through before deciding on a direction of travel.

How do you ensure your organization and its activities are aligned with your core values?

It's difficult to turn the big cogs of a wheel in an organisation of my size (LHR has 8,000 employees), but if you display the right characteristics, stick to your beliefs of honesty and transparency then hopefully, over time, this will rub off on to others.

What are you doing to ensure you continue to grow and develop as a leader?

Continue to read reports, keep up with legislation, attend seminars and workshops for CPD as well as internal "soft skill" training. Also keep up with my own operational training so I am as prepared as can be when the next major incident happens.

* * *

On 28 November 2008, a Boeing 777-200ER suffered an in- flight engine rollback; an investigation by the NTSB was initiated with...the AAIB.

Whilst on approach to London (Heathrow) from Beijing, China, at 720 feet agl, the right engine...ceased responding to autothrottle commands for increased power and instead the power reduced to 1.03 Engine Pressure Ratio (EPR). Seven seconds later the left engine power reduced to 1.02 EPR. This reduction led to a loss of airspeed and the aircraft touching down some 330 m short of the paved surface of Runway 27L at London Heathrow. The investigation identified that the reduction in thrust was due to restricted fuel flow to both engines.[1]

— AIR ACCIDENTS INVESTIGATION BRANCH

What lessons did your department learn?

That continued training, however repetitive, is the key to ensuring a fast and competent response of the crews and Incident Commanders.

That the inter-liaison relationships and Incident Command processes could have been better (between municipal FS and ARFF).

That equipment to enter an aircraft could be improved, as at the time, we only had an Arial Ladder Platform (ALP).

That our Crash call-in system needed updating; as this was a system of manually phoning off duty firefighters to attend as extra cover.

What changes were made?

We introduced a yearly program of liaison training with London Fire Brigade to address ways of working, knowledge of aircraft and ARFF appliances, regular joint exercises, and aligned ourselves to the National Incident Command System so that everyone was working to the same processes and principles. This included procedures, terminology, tabards etc.

We procured a set of Rosenbauer Rescue Stairs as a better means for ARFF crews being able to enter an aircraft in an emergency and also aid swift passenger egress (highest point being an A380 upper deck door 8.3m).

We upgraded to a digital call-in system that texts and calls firefighters off duty, taking away the manual slow process.

. . .

Had your training and experience to this point adequately prepared you for this incident? How, why, or why not?

Yes our training program ensured that our deployment to the scene and subsequent actions were suitable and sufficient

What impact has this had on your career and leadership?

It has made me value the need to ensure there is a team spirit amongst the crews, together with a good training plan, to ensure confidence and competence for events like these which are statistically very rare. Being prepared at all levels is key to success.

What advice would you give to prepare someone for responding to an incident of this magnitude?

To trust your years of training and experience in that it will prepare you as best as possible for the real thing. Realistic, regular training putting enough pressure on crews and Incident Commanders is key. Therefore, a robust training plan, regular maintenance of appliances and fire service equipment, and good well-drilled procedures and emergency plan should prepare you as best as possible. It will never replicate totally but will serve you as best as practically possible.

34

JOSEPH MARINO, CHIEF OF DEPARTMENT, PORT AUTHORITY OF NEW YORK & NEW JERSEY, AIRCRAFT RESCUE AND FIREFIGHTING

How did you get to your position? What path did your career follow?

While serving as a Fire Chief at Long Island MacArthur Airport, I became aware of significant changes within the ranks of the Port Authority Aircraft Rescue Firefighting Unit (ARFF). The agency was heading in a new direction with respect to leadership and I thought it would be a perfect opportunity for me to explore. In 2014, I was hired by the Port Authority as the LaGuardia Airport ARFF Commanding Officer. Soon after, I was promoted to Deputy Fire Chief. In 2019, I was promoted to Chief of Department.

What is your advice for the newly promoted chief officer?

Whatever rank you are, do not be afraid to embrace your weaknesses as a chief officer, and solicit the advice, consult, and assistance of those who can help you to have career success. People in our profession are often programmed to hold an image of being the one with all the answers. After all, that is why you got promoted right? Wrong! I have been a Fire Chief in both a career fire department, as well as, a volunteer fire department. My success in both fields was due in part to my fellow chief officers. I recognize I have strengths and weaknesses. It is a part of being human. Putting pride and bravado aside, I sought out advice and consult from others. I am very much proud of my success. I did not get where I am on my own. So, as long as you are entrusted to be a chief officer or any position in the fire service, you owe it to yourself and your fellow firefighters to do whatever is necessary to make you a better supervisor.

What was the most impactful call/emergency you have been on? Why?

As a new Lieutenant, I arrived first due to a working fire in a private dwelling. As the engine crew was making a push in a heavily involved first floor, the water stream passed me and knocked my helmet off. The heat I felt was something I never felt before. Lieutenant Farrell from another engine company somehow found my helmet and placed it back on my head, of course baptizing me with hot water that had gathered in the helmet from water being thrown in the fire room. I was ahead of the hand-line and once again the water stream knocked my helmet off exposing my head again to a superheated environment. Lieutenant Farrell was relatively near me but was unable to locate my helmet. Lieutenant Farrell put his gloved hand in front of the hand-line stream, soaking his glove. He immediately placed his water soaked glove on my head, cooling it down. My helmet was located soon after and the rest of the job was handled with no events. Why was it so impactful? Well first, did my own vanity prevent me from using the chin straps on my helmet? Yes! Had I suffered serious burns, I would have nobody to blame but myself. No excuse. Second, why did I not think of cooling off my own head with a waterlogged glove? It is common sense right? No, it is not common sense. The move was a product of a fire officer truly prepared to make an unconventional move. You can't be taught everything you will be presented with while in the fire service. You have to be thinking every second, expect the unexpected, adapt and overcome. This was a humbling experience for me because I know that as much as I thought I was prepared to be a boss, I knew I had so much to still learn.

What actions, behaviors, or thought patterns lead to leadership failure?

This is a tough question because all of us fire service leaders are not without fault, to a degree. The successful leaders I surround myself with are competent, empathetic, dedicated, humble and respectful. If you are none of those, maybe all, I would expect that you will have a tough time being a strong leader

What is the top trait or characteristic that you believe every chief officer must possess?

In the words of Tim McGraw, "always stay humble and kind"! A chief officer can be viewed as the "big man on campus". Stay humble knowing that. Stay kind to the probie who is terrified to even walk past you.

What is a habit or routine that you have and how does it help you persevere?

Stay grounded and learn something new every day. Even if it has nothing to do with the fire service.

What have been the key factors to your success and why?

I have been fortunate to have a successful career in the fire service. I came from a humble small town fire department and became the leader of the fire department in the world's busiest airport system. A key factor to my success is always keeping in touch with that small town fire department that I served in and reminding them that they launched me to my success.

. . .

What are the most important decisions you make as a leader of your organization?

Discipline. We have rules in the fire service. As a fire service leader, you will inevitably be presented with a situation where you have to discipline a friend. Discipline must be handed down equally no matter what fallout that could occur.

How do you ensure your organization and its activities are aligned with your core values?

My core values are important to me. If there was an instance where my core values are not consistent with my organizations then I will absolutely have discussions with my superiors. I will always respectfully maintain my core values.

What are you doing to ensure you continue to grow and develop as a leader?

Never sitting back and saying I have a quiet day today. There is always something to learn, someone to meet, a question to be asked, an email to be answered, a better way to do business, a check-in with a colleague who would appreciate it. The list goes on. I recognize that I reached a position that many would dream of having. Taking this for granted for even one second, in my mind, is a failure to the respect my position is deserving of.

35

MARK HUETTER, BATTALION CHIEF, NASA/KENNEDY SPACE CENTER

How did you get to your position? What path did your career follow?

I began in the Air Force with drive to be at the top from day one.

What is your advice for the newly promoted chief officer?

Don't forget where you came from, from infant to now, from location to now, from schools to now, from life situations (good and bad) to now, and from the beginning of your career to now. Take all that you have learned, good and bad, and let it mold you and define you and help you become the best CHIEF you can be.

What was the most impactful call/emergency you have been on? Why?

All of them. Think about it, we are humans, we are in a profession that deals with affecting an individual 100% of the time, whether it be financially, emotionally, physically, mentally. Every call from false alarms to mass casualty.

. . .

What actions, behaviors, or thought patterns lead to leadership failure?

Failure to be a leader.

What is the top trait or characteristic that you believe every chief officer must possess?

Honor.

What is a habit or routine that you have and how does it help you persevere?

Prayer. Remembering that I serve a Greater Power and knowing I have an amazing and beautiful family.

What have been the key factors to your success and why?

Faith, family, and drive.

What are the most important decisions you make as a leader of your organization?

The most important decisions I make as a leader are the ones that deal with the safety of all my men to assure they return to their families at the end of a shift.

How do you ensure your organization and its activities are aligned with your core values?

Communication.

What are you doing to ensure you continue to grow and develop as a leader?

I continue to grow and develop.

ALLEN WARD, REGIONAL DIRECTOR/COMMAND FIRE CHIEF, RURAL/METRO FIRE DEPT., INC.

How did you get to your position? What path did your career follow?

Volunteer Firefighter - 1978 - 1988

Military Firefighter - 1988 - 2004

ARFF Firefighter (Port Columbus Int'l Airport) - Jan 2001 - Sep 2001

Fire Captain (Port Columbus Int'l Airport) - Oct 2001 – Nov 2002

ARFF/EMS/Security Chief (Port Columbus Int'l Airport) - Dec 2002 - Sep 2006

North Regional Command Fire Chief, Specialty Fire Operations - Oct 2006 – Apr 2010

Operations Command Chief, Specialty Fire Operations - May 2010 – Dec 2012

Director Specialty Fire Operations - Jan 2013 – Present

What is your advice for the newly promoted chief officer?

Listen to your subordinates before making a final decision.

. . .

What was the most impactful call/emergency you have been on? Why?

As a volunteer Assistant Chief, I responded to a structure call and during the walk through after extinguishment we found a deceased individual sitting in a recliner that had melted. This was an individual that I had graduated from high school with, and this was my first time dealing with a deceased individual.

What actions, behaviors, or thought patterns lead to leadership failure?

I believe that any action, behavior, or thought pattern could lead to leadership failure if the leader does not maintain an eye on the pulse of his department.

What is the top trait or characteristic that you believe every chief officer must possess?

Ability to listen.

What is a habit or routine that you have and how does it help you persevere?

Making sure that I meet all deadlines regardless of the time I have to spend to complete that task.

What have been the key factors to your success and why?

Doing what is expected of me, without objection.

What are the most important decisions you make as a leader of your organization?

Providing and managing the overall direction of a twenty-four station, thirteen state division.

. . .

How do you ensure your organization and its activities are aligned with your core values?

By remaining fixed on my core values, even as the business strategies and tactics shift to adapt to changes in the industry, or new levels of growth.

What are you doing to ensure you continue to grow and develop as a leader?

Reading industry magazines on technical procedures, continuing education.

III

LEADING ON

37

HALL OF MENTORS

THE FINAL QUESTION asked of each of the ARFF leaders and officers for this book was, "Can you name a person who has had a tremendous impact on you as a leader? Someone who has been a mentor?" These individuals took time to train, teach, and nurture those under them or equal to them. The majority of mentors listed were fathers and grandfathers, then followed by officers worked with and for. Here we want to honor and acknowledge those individuals who were named.

Fathers and Grandfathers · Raymond Burchill, Assistant Chief · JP Apon, General · Peter Caspersen, Chief · Carlos Sanchez, Captain · Jim Nilo, ARFF Chief · Jack Kreckie, ARFF Chief · Bill Stewart, ARFF Chief · Greg Gill · Ray Bishop, Colonel · Wayne Dukes, Fire Chief · Matt Mauer, Battalion Chief · Shane Clark, Fire Chief · Bruce Niven, Colonel · Jane Sherlock, Director · General Zumstein · Mick Garner, Instructor · Chris Farnaby, Instructor · Alan Brunacini, Fire Chief · Dennis Compton, Fire Chief · Wayne Sandford, Fire Chief · George Klug, VP of Safety · Randy Krause, Chief · Joseph Wright, Fire Scientist · Edmund T. Burke, Captain · Richard Cook, Firefighter · Keith Wheatley, Duty Officer · Curtis Boehmer, Wing Commander · D. Provan · K. Leahy · L. Freer · Lonnie Clark

38

RESOURCES FOR LEADERS

THE BELOW BOOKS and organizations were the most frequently cited by the ARFF leaders interviewed. These are their recommendations for continued education and professional development.

- *Holy Bible*
- *First In, Last Out,* John Salka
- *It's Your Ship,* Michael Abrashoff
- *Pride and Ownership,* Rick Lasky
- *Essentials of Fire Department Customer Service,* Alan Brunacini
- *From Buddy to Boss,* Chase Sargent
- *The Fifth Discipline,* Peter Senge
- *Extreme Ownership,* Jocko Willink
- *The Dichotomy of Leadership,* Jocko Willink and Leif Babin

- ARFF Working Group membership
- National Fire Protection Association (NFPA) codes and publications
- National Transportation Safety Board (NTSB) Accident Reports

AFTERWORD

Several years ago Malcolm Gladwell, in his book *Outliers*, brought attention to the "ten-thousand-hour rule". This is the idea that it takes ten-thousand hours of concentrated effort to master any task, skill, or ability. This has since been a source of controversy for psychologist and performance professionals. However, Gladwell clarified the "ten-thousand-hour rule" in a piece for The New Yorker:

> "No one succeeds at a high level without innate talent, I wrote: "achievement is talent plus preparation." But the ten-thousand-hour research reminds us that "the closer psychologists look at the careers of the gifted, the smaller the role innate talent seems to play and the bigger the role preparation seems to play."[1]

This book has been written to help close the "ten-thousand-hour" gap and to serve as a component of "preparation" for leadership. This is a tool for the ambitious and forward moving ARFF professional to gaining the knowledge and skills needed to achieve their leadership goals. Hopefully, by reading and learning from the years of experience shared here, the gap to achieving a leadership position can be decreased.

Within the next five to ten years, another volume of this type will need to be researched, compiled, and written. We will have a new group of ARFF leaders and retirees from which much knowledge can be garnered. It is my sincere hope that this book will be the start of many other volumes that will focus their attention on the unique challenges faced by the ARFF industry.

NOTES

INTRODUCTION

1. *What is KM? Knowledge Management Explained.* (2018, January 15). KMWorld. https://www.kmworld.com/About/What_is_Knowledge_Management
2. *The Consultative Approach to Fire Protection Problems.* (n.d.). TheCodeCoach.Com. Retrieved October 3, 2021, from https://bit.ly/Consultative_Approach
3. *After Action Review.* (2015, January 29). Everyone Goes Home. https://www.everyonegoeshome.com/16-initiatives/13-psychological-support/action-review/

1. LEADERSHIP LESSONS

1. *Oxford Languages and Google - English|Oxford Languages.* (2021, April 9). Oxford Languages. https://languages.oup.com/google-dictionary-en/
2. *Enhancing Fire-Rescue Human Capital: Trust in the Fire Service.* (n.d.). International Association of Fire Chiefs. https://www.iafc.org/iCHIEFS/iCHIEFS-article/enhancing-fire-rescue-human-capital-trust-in-the-fire-service
3. John Maxwell, *The 360° Leader: Developing Your Influence from Anywhere in the Organization,* https://amzn.to/3voIhFb
4. Brene Brown, *Dare to Lead,* https://amzn.to/3iF6pkb
5. Brown, B. (2021, August 28). *Clear is Kind. Unclear is Unkind - Brene Brown.* https://brenebrown.com/blog/2018/10/15/clear-is-kind-unclear-is-unkind/#close-popup
6. *Sun Tzu and the Art of Fireground Leadership,* https://amzn.to/3A186NL
7. *Honest Meaning|Best 18 Definitions of Honest.* (n.d). YourDictionary.com. https://www.yourdictionary.com/honest
8. Abrams, A. (2018, June 20). *Yes, Imposter Syndrome Is Real. Here's How to Deal With It.* Time. https://time.com/5312483/how-to-deal-with-impostor-syndrome/
9. Posts, V.M. (2020, November 5). *Embracing Imposter Syndrome.* Jonah Baer. https://jonahbaer.com/2020/11/05/embracing-imposter-syndrome/
10. Godin, S. (2021, February 28). *The Practice: Ship creative work.* Seth's Blog. https://seths.blog/ThePractice/
11. *Integrity Meaning|Best 17 Definitions of Integrity.* (n.d). YourDictionary.com. https://www.yourdictionary.com/integrity
12. J. (2013, March 30). *Stephen Covey on How Experience is Overrated.* Sources of Insight. https://sourcesofinsight.com/stephen-covey-on-how-experience-is-overrated/
13. Foskett, J. (n.d.). *The Impact of Alan Brunacini: what all firefighters should know about 'America's fire chief.'* FireRescue1. https://www.firerescue1.com/leadership/arti-

cles/the-impact-of-alan-brunacini-what-all-firefighters-should-know-about-americas-fire-chief-MSfWicvv5wuyjKPZ/
14. *Matthew 7:12 (HCSB).* (n.d.). Bible Gateway. https://www.biblegateway.com/passage/?search=Matthew%207:12&version=HCSB

9. SCOTT LANTER, A.A.E., DIRECTOR OF PUBLIC SAFETY AND OPERATIONS, BLUE GRASS AIRPORT

1. NTSB Accident Report on Flight 5191. https://www.ntsb.gov/investigations/AccidentReports/Reports/AAR0705.pdf

12. WILLIAM MAJOR, FIRE CHIEF, BUFFALO NIAGARA INTERNATIONAL AIRPORT

1. NTSB Accident Report on Flight 3407. https://www.ntsb.gov/investigations/AccidentReports/Reports/AAR1001.pdf

18. ALVIN LEE, CHIEF, AIRPORT EMERGENCY SERVICE, CHANGI AIRPORT GROUP

1. Ministry of Transport Flight 368 Accident Report. https://www.mot.gov.sg/docs/default-source/about-mot/investigation-report/b773er-(9v-swb)-engine-fire-27-jun-16-final-report.pdf

33. GARY BARTHRAM, CHIEF FIRE OFFICER, AIRPORT FIRE & RESCUE SERVICE, LONDON HEATHROW AIRPORT

1. AAIB Final Accident Report on British Airways Boeing 777-236ER. https://assets.publishing.service.gov.uk/media/551d5725e5274a142e00047f/Summary_AAR_1-2010_Boeing_777-236ER__G-YMMM_02-10.pdf

AFTERWORD

1. Gladwell, M. (2013, August 21). *Complexity and the Ten-Thousand-Hour Rule.* The New Yorker. https://www.newyorker.com/sports/sporting-scene/complexity-and-the-ten-thousand-hour-rule

ACKNOWLEDGMENTS

I would like to acknowledge and express my sincere depth of gratitude to every individual who contributed to this book, and those departments who approved their participation. Without your willingness to take the time to participate this work would not have been possible. I am humbled and honored that you chose to be part. Thank you so much for sharing your wisdom, knowledge, and experience with me and these readers.

ABOUT THE AUTHOR

Aaron has more than fifteen years of experience in the fire service and ARFF industry. He started his career as a firefighter, then progressed through the ranks to the position of Fire Marshal. He is actively involved in the codes and standards development process, specifically those codes related to aviation and ARFF response. Aaron is currently the Chief Fire Strategist specializing in innovative fire protection solutions for emerging technologies. This is his sixth book written for the fire service.

Website: www.aaronj.org
Email: thecodecoach@gmail.com
LinkedIn: https://www.linkedin.com/in/baaronj/
Twitter: https://twitter.com/thecodecoach
Instagram: https://www.instagram.com/b_aaron_j/

ALSO BY AARON JOHNSON

Sun Tzu and the Art of Fireground Leadership

Fire Protection for Hangar Design: A Guide for Compliance in Aviation Facilities

Risk Assessment Guide for Aviation Facilities

Fire Prevention Blueprint: Seven Disciplines for Building Effective Fire Prevention Organizations

Fire Sprinkler Design: An Illustrated Guide for AHJ's

www.ingramcontent.com/pod-product-compliance
Lightning Source LLC
Chambersburg PA
CBHW052315220526
45472CB00001B/134